ELECTRONIC SYSTEMS

ELECTRONIC SYSTEMS

for Radio, Television and Electronic Mechanics

Rhys Lewis, B.Sc.Tech., C.Eng., M.I.E.E.

Head of Electronic and Radio Engineering Department,
Riversdale College of Technology,
Liverpool

First edition 1973
Reprinted 1976, 1979 (with corrections)

Published by
THE MACMILLAN PRESS LTD
London and Basingstoke
Associated companies in Delhi Dublin
Hong Kong Johannesburg Lagos Melbourne
New York Singapore and Tokyo

ISBN 0 333 13712 4

Printed in Great Britain by
UNWIN BROTHERS LIMITED
The Gresham Press, Old Woking, Surrey
A member of the Staples Printing Group

Preface

This book is written specifically for the City and Guilds of London Institute course 222 (formerly 433) for radio, television and electronics mechanics. It covers the complete syllabus of the subject 'Electronic Systems' (both parts I and II) in one volume. To a large extent the book also covers the systems content of CGLI course 932 (formerly 969) for Telecommunications mechanics, and should prove useful to students on CGLI course 270 (formerly 434) for radio, television and electronics technicians.

The layout of the text is in accordance with the syllabus layout of the 222 course. It commences with an explanation of signals, system subunits (amplifiers, oscillators etc.) and system block diagrams. Detailed circuits are examined throughout the next seven chapters, and the final chapter is devoted to faults (causes and location), environmental considerations, and reliability. Appendixes cover resistor colour coding, matching, valve and transistor parameters, and multiple units; and the book concludes with fifty self-test questions of the kind used by the CGLI in examinations for the courses mentioned.

The approach to the theory presented is strictly non-mathematical and physical explanations have been used throughout. Numerous diagrams are included, many of which indicate typical values of components found in practical circuits. In studying these circuits the reader should bear in mind that component values may differ considerably from those indicated, as in many applications the values depend on the characteristics of the particular valve or transistor with which these components are used. Another point to note is that though components are labelled in the traditional manner, reference is made in appendix 1 to newer methods (as suggested in BS 1852) which some manufacturers are now adopting.

I would like to offer my most sincere thanks to Mrs. H. M. Morgan for her efforts in typing and preparing the manuscript, particularly as she stepped into the breach at a very late stage in the proceedings, thereby enabling the text to reach the publishers by the proposed date. Last, but by no means least, I would, as ever, like to thank my wife for her continual encouragement and patience during many hours devoted to writing the book.

RHYS LEWIS
Openshaw, Manchester
September 1972

Contents

1 Electronic Signals

Electronic systems range from domestic radios, television and record-players to the sophisticated computing, control and recording systems used in industry and transport (including military and aerospace vehicles). The purpose of any electronic system is to convey intelligence from one point to another; the word intelligence in this application meaning information and including words, music, pictures and numerical data. Such intelligence is conveyed from source to receiver by means of *electronic signals.*

1.1. D.C. Signals

To illustrate the nature of an electronic signal and in particular a d.c. (direct current) signal, consider a simple electrical circuit made up of a d.c. power supply, a d.c. electric motor and a switch, as shown in figure 1.1a. When the switch is closed voltage is applied by the supply (source) across the motor (receiver), current flows and the motor runs. When the switch is open there is no path for current flow, the voltage across the motor is zero and the motor does not run. If a graph plotting voltage against time is drawn as in figure 1.1b, the intervals during which the motor runs or does not run are clearly shown. The intelligence sent in this case from source to receiver consists of the two instructions 'stop' and 'run'. The voltage or current is the signal by which the intelligence is carried through the system. Once the circuit is closed voltage is applied in one direction only, current flows in one direction only and the signal is called a direct current or d.c. signal.

If a double-pole, double-throw switch is substituted for the single-pole single-throw switch of figure 1.1a, the circuit becomes as in figure 1.2a and it is now possible to reverse the direction of current flow and thus the direction of rotation of the motor. Three instructions can now be conveyed through the system: 'run in one direction', 'run in the reverse direction' and 'stop'. The graph of voltage against time now becomes that shown in figure 1.2b.

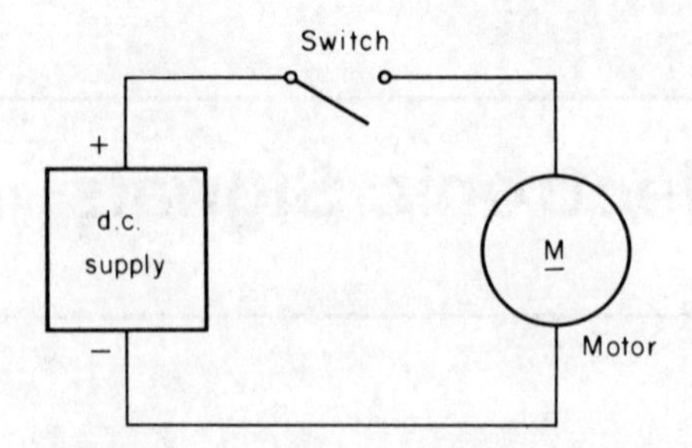

(a) Simple electrical circuit using single-pole, single-throw switch and d.c. motor.

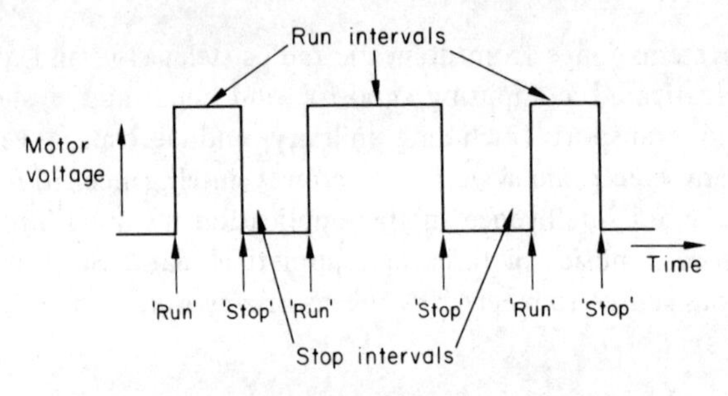

(b) Graph of voltage received by motor in circuit of figure 1.1 (a)

figure 1.1 The unidirectional d.c. signal

The system is still using a d.c. signal, as the instruction is sent from source to receiver by a current that flows in only one direction (the particular direction being determined by the side to which the switch is closed).

Both d.c. systems so far considered are relatively simple, in that the motor either runs at constant speed once the switch is closed, or is stationary when the switch is open. Figure 1.3a shows a circuit in which the voltage applied to the motor is both variable and reversible and the 'run' instruction becomes more detailed, as it is now possible to choose the running speed by changing the magnitude of the applied voltage. The graph of voltage against time is shown in figure 1.3b. The left-hand part of the graph shows the effect of switching on and off with the variable supply preset at different levels; the right-hand part shows the effect of changing the voltage supplied to the motor with the switch closed while the variation is being made. This system can transmit the instructions 'run in one direction at a particular speed', 'run in the other direction at a particular speed' and 'stop'. Once running at a speed other than maximum the instructions 'go faster' and 'go slower' may also be transmitted from source to receiver.

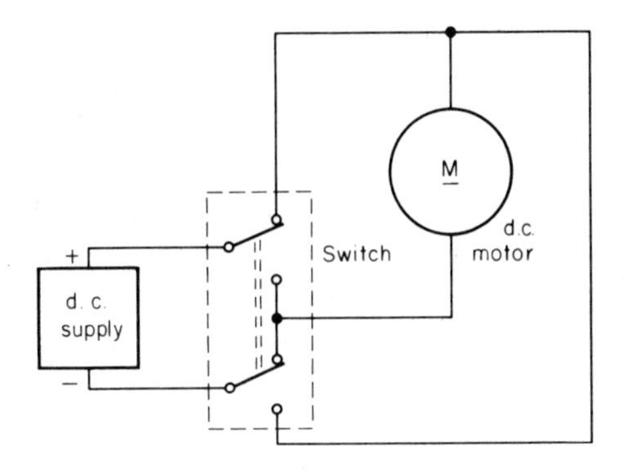

(a) Motor circuit with reversing facility

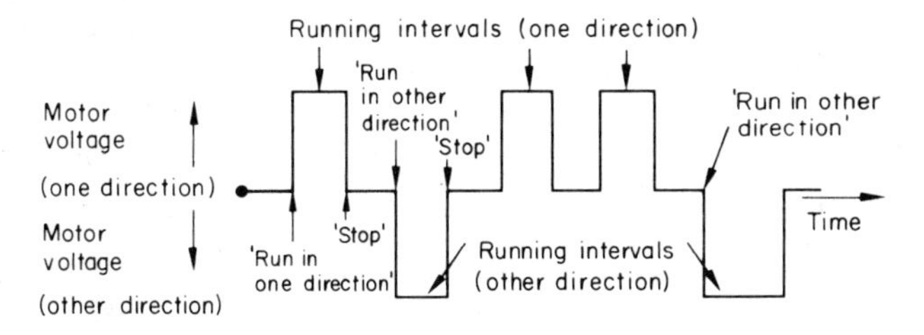

Motor is stopped when graph lies along horizontal axis

(b) Voltage versus time graph
for circuit of figure 1.2 (a)

figure 1.2 The two-directional d.c. signal

Control of motors by d.c. signals is applied to very large high-power machines (for example, in steel mills), to medium-size machines (for example, for automatic door opening equipment or conveyor belt systems) and to very small machines (for example, pen recorders: where a pen placed on a moving paper roll records changes in some particular quantity such as temperature or pressure). D.C. signals are also used in such equipment as light-exposure meters (in which the magnitude of a direct current is varied by the intensity of illumination in the control region, the signal being then used to determine camera settings) and in code transmission systems that use the rise and fall of a voltage level to carry verbal intelligence in code form (for example, Morse, as shown in figure 1.4). Computers and other forms of logic circuitry also use d.c. signal transmission.

There are two particular disadvantages of using d.c. signals: firstly, there is difficulty in changing the signal to higher or lower values of voltage or current; and secondly, a system employing only d.c. signals always requires

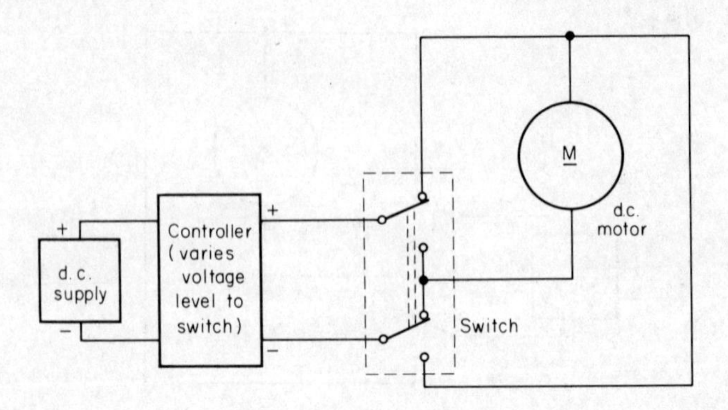

(a) Motor circuit with speed adjustment and reversing facility

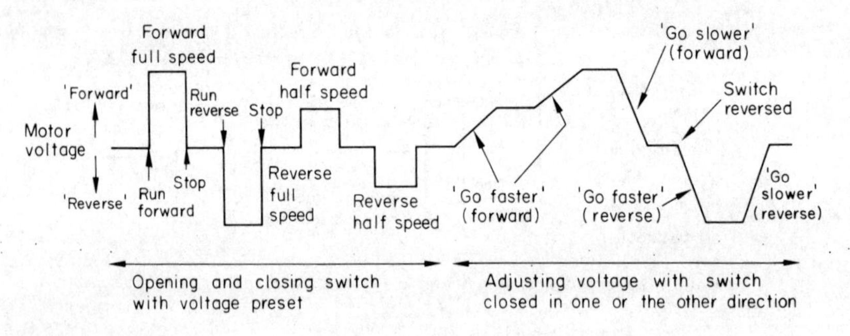

(b) Graph of voltage against time for circuit of figure 1.3 (a)

figure 1.3 Two-directional d.c. signal with variable voltage

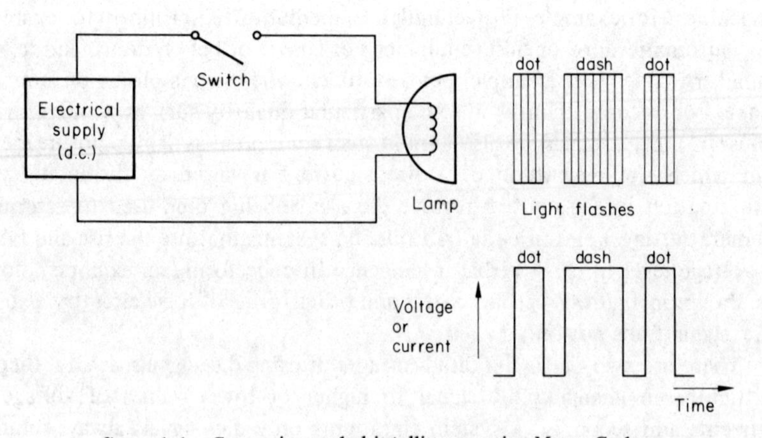

figure 1.4 Conveying verbal intelligence using Morse Code

connecting wires between source and receiver. Thus while d.c. signals cannot by themselves be transmitted without wires, alternating current (a.c.) signals can.

1.2. A.C. Signals

An *alternating current* is one that periodically changes its direction of flow within a circuit. If such a current is used to transmit intelligence it is called an a.c. signal. A typical graph of an a.c. signal, showing current plotted against time, appears in figure 1.5; the same shape would result if the voltage that

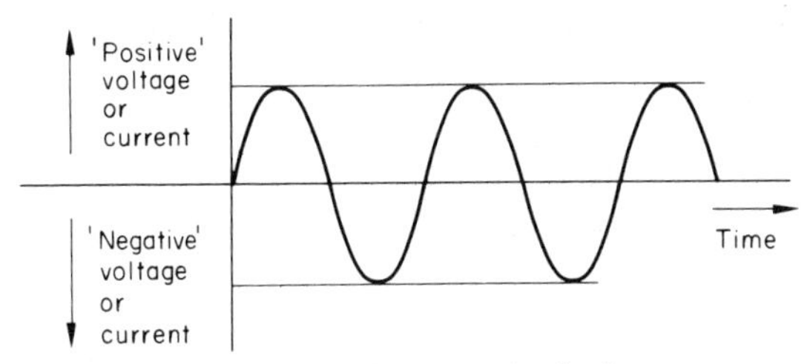

figure 1.5 An a.c. signal of sine-wave form (graph of voltage or current against time)

caused the current were plotted against time, because the voltage alternates in the same manner (although not necessarily at exactly the same time). The nature of this particular variation, called a *sine wave,* and certain associated definitions, will be examined later. The changing nature of an alternating current or voltage enables it to be easily changed to larger or smaller values by use of a device known as a *transformer.* Also, provided the rate of change of current or voltage is fast enough, it is possible to transmit an a.c. signal without using connecting wires between source and receiver (because of the existence of electromagnetic waves, which are discussed in more detail in the next article).

It might be contended that figures 1.2b and 1.3b show a current changing direction just as does figure 1.5. The common characteristic must not, however, be allowed to cause confusion between the natures of the two signals. Figures 1.2b and 1.3b show d.c. signals and the change in current direction causes a change in the transmitted intelligence: the instruction changes from 'run in one direction' to 'run in the other direction'. Figure 1.5 shows an a.c. signal and the change in current direction does not imply a change in the intelligence transmitted. To illustrate this point consider figure 1.6 which shows the graphs of various alternating currents having different rates of change. If these currents were fed singly into a sound-reproducing device such as a loudspeaker, a single tone would be heard each time; the

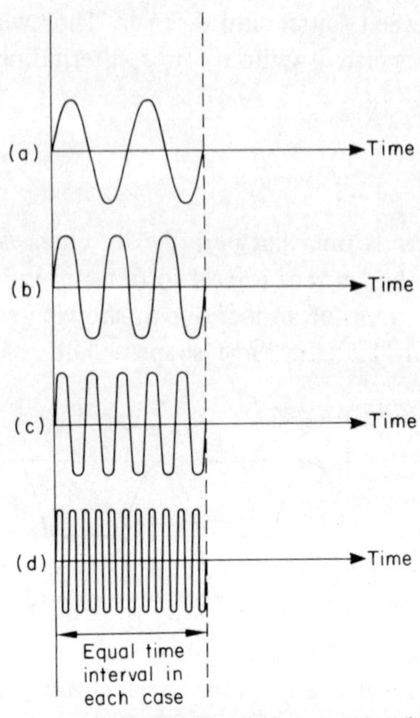

figure 1.6 a.c. signals having different rates of change

current of figure 1.6a giving the lowest tone and the current of figure 1.6d giving the highest tone. The intelligence transmitted each time (that is, the single tone) remains the same even though the current alternates. Indeed, as will be discussed shortly, it is the fact that the signal is an alternating current which produces the tone. A direct current similarly fed to a loudspeaker would produce no tone because it does not change periodically.

A.C. signals are widely used for intelligence transmission. The best known systems are probably those of radio, television and sound reproduction from tapes or records; but there are many other applications, one of which is the transmission of numerical or coded data between a central control and remote vehicles such as aircraft, space research rockets and satellites. To begin a deeper investigation of the nature and uses of a.c. signal transmission it is first necessary to examine in more detail the different types of a.c. signal, devoting particular attention to electromagnetic waves and the electromagnetic spectrum.

1.3. The Electromagnetic Spectrum

The only kind of a.c. signal so far considered is that of an electric current changing in magnitude and direction within a physical electric circuit. A circuit of some kind (that is, conductive wires and components) is necessary to carry the current and an a.c. signal of this form is therefore restricted in the same way as a d.c. signal. However, it is found that if an alternating current changes fast enough, it is possible to transfer energy (and thus intelligence) from one circuit to another that is not linked by wires to the first. This is due to the setting up of electromagnetic energy waves which radiate from the one circuit and are received by the other.

To obtain a picture of energy transmission by wave motion consider the following examples: the first shows the creation of a wave and how it travels; the second shows how such a wave carries energy. Imagine a rope tied at one end to a wall, the other end being held in the hand. If the hand is now moved up and down, as shown in figure 1.7, a wave travels down the rope towards the fixed end. Each part of the rope remains in the same vertical plane, but the wave travels in the horizontal plane. The second example is that of a stone dropped into a pond. Waves radiate radially from the point of impact, the propagation taking place in much the same way as in the previous example. Any object floating nearby will absorb energy and rise and fall. This example illustrates the transfer of energy from the stone to the floating object via the water wave.

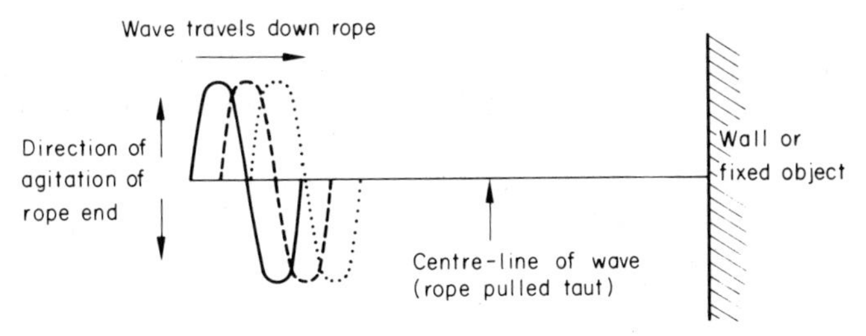

figure 1.7 Generating a rope wave

Whenever a current is alternating within a conductor an electric field is set up around the conductor. This field builds up in strength and then collapses as the current changes. In addition, a magnetic field is also set up, at right angles to the electric field, and the magnetic field strength also increases and reduces alternately. The combination of the changing electric and magnetic fields sets up an electromagnetic wave which radiates from the conductor carrying the current in much the same way as the water wave discussed earlier. The distance covered by the radiation and the strength of the electromagnetic wave generated is determined (among other things) by the conductor length, by the rate of change and magnitude of the alternating current and by the associated voltage. The topic of propagation is dealt with more fully in chapter 10.

It should be noted that an electromagnetic wave is not exactly the same as those discussed earlier, for both the water wave and rope wave travel within a medium (that is, water, or rope, respectively) whereas an electromagnetic wave can travel within a vacuum, because the electromagnetic wave is a rise and fall in the strength of electric and magnetic fields and these do not require matter for their existence. However, the alternate strengthening and weakening of the fields does result in a wave pattern that travels outwards from the source and thus can be used for the transmission of energy and intelligence.

The wave pattern of an electromagnetic wave is similar to the one already given for an alternating current contained within a circuit (that is, a sine wave); but whereas the sine wave depicting an alternating current represents current magnitude plotted against time, the wave shape of an electromagnetic wave can be regarded as a plot of field strength (as determined by the combination of electric and magnetic fields that produced the wave) against distance travelled. The two waveforms in transmission are, of course, closely related, as it is the alternating current within the circuit which produces the electromagnetic wave outside the circuit.

There are several important characteristics of a wave, certain of which are common to both the a.c. sine wave and the electromagnetic sine wave and certain of which are more commonly used with one or other of the two waves as explained below. The important characteristics are amplitude, wavelength, periodic time and propagation velocity. (See figure 1.8.)

The amplitude is the peak value reached in either direction from the centre of the wave. When the sine wave represents an alternating current or voltage other values of magnitude are also important. These are the average value and

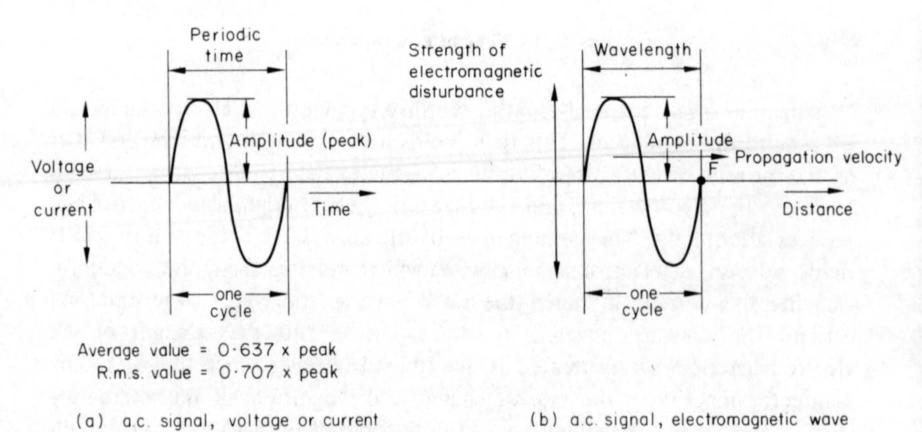

figure 1.8 Types of a.c. signal wave

the root-mean-square or r.m.s. value. The average value of a complete cycle is, of course, zero as there are equal excursions above and below the centre line. For a sine wave the average value of half a cycle is the one normally taken and it can be shown that it is in fact equal to 0.637 of the peak value or amplitude. The average value is the one that determines the deflection of certain instruments such as those of the moving coil variety which are used for current or voltage measurement. The r.m.s. value is the value of the equivalent direct current or voltage to do the same work (that is, provide the same energy) under the same circumstances as the alternating current or voltage. It can be shown that the r.m.s. value of a sine wave is equal to 0.707 of the peak value. The r.m.s. value is the one most commonly used when referring to alternating currents or voltages—the mains voltage in the United Kingdom, for example, being given as 240 volts has an r.m.s. value of 240 V and thus a peak value of 240/0.707, that is, 339.4 V. The average value of the mains voltage over one cycle would then be 339.4 x 0.637, that is, 216.2 V.

The *frequency* of a wave is the number of cycles per second, where a cycle is a complete alternation from zero to maximum value in one direction, back through zero to maximum value in the other direction, and finally back to zero (see figure 1.8). One cycle per second is called one *hertz* (abbreviated Hz) and the multiple units include the kilohertz (kHz) which is 1000 cycles/second, the megahertz (MHz) which is 1 000 000 cycles/second and the gigahertz (GHz) which is one thousand million cycles/second. The frequency of the a.c. mains supply in the UK is 50 Hz and in the USA and Canada is 60 Hz. The frequency of a wave is extremely important as to a great extent it determines the use to which the wave can be put.

The *periodic time* is the time taken for one cycle. A moment's thought will show that the periodic time is the reciprocal of the frequency, for if the frequency is f cycles per second (hertz), then the time taken for one cycle must be $1/f$ seconds. If the wave is of an alternating current or voltage then the periodic time can be shown directly on the graph as in figure 1.8a. It cannot be shown directly on the graph of an electromagnetic wave, because distance and not time is plotted along the horizontal axis.

The *wavelength* is a term used particularly when discussing electromagnetic waves and is the distance occupied by one cycle. (See figure 1.8b.)

The *propagation velocity* is again a term used in connection with electromagnetic waves and is the velocity with which a wave travels. In fact, all electromagnetic waves travel at the velocity of light, namely 300 million metres per second (about 186 thousand miles per second). The wavefront of an electromagnetic wave (that is, the point F in figure 1.8b) travels a distance equal to one wavelength in the time taken for one cycle. In one second the number of cycles formed is known as the frequency. Thus the distance travelled by the wavefront in one second is equal to the product of frequency and wavelength. The distance travelled in one second is, of course, the propagation velocity so that propagation velocity = frequency x wavelength; and, since the propagation velocity is constant, frequency is proportional to 1/wavelength. For all electromagnetic waves we see therefore that as the frequency is increased the wavelength is reduced; that is, long waves are at low frequencies, short waves are at high frequencies.

table 1.1 The Electromagnetic Spectrum

	Frequency in Hz		Wavelength in m	
	3.0×10^{22}		10^{-14}	
	3.0×10^{20}		10^{-12}	Cosmic rays
Gamma rays	5×10^{19}		6×10^{-12}	
	1.5×10^{18}		2×10^{-10}	
	2.5×10^{16}		1.2×10^{-8}	X-rays
Ultra violet rays	3×10^{15}		10^{-7}	
	7.5×10^{14}		4×10^{-7}	
	3.75×10^{14}		8×10^{-7}	Light waves
Infra red (heat) waves	3×10^{12}		10^{-4}	
	7.5×10^{11}	Hertzian (radio) waves	4×10^{-4}	
Radar TV VHF Radio Bands	8.9×10^{8}		3.37×10^{-1}	These are sample frequencies within this band
	4.7×10^{8}		6.38×10^{-1}	
	5.4×10^{7}		5.55	
Short wave	1.6×10^{6}		187.5	
Medium long waves	2×10^{4}		1.5×10^{4}	Limit of human ear
	20 Hz		1.5×10^{7}	

Notes. Details of actual bands used in the UK and elsewhere for Radar, radio, short and medium-long waves are given in subsequent chapters
10^{12} means 1 000 000 000 000 (that is, 1 followed by 12 zeros)
10^{-12} means 0.000 000 000 001, that is, zero followed by 11 (12 – 1) zeros; in other words there are (12 – 1) zeros following the decimal point

The complete range of electromagnetic waves varies from very low to very high frequencies (very long to very short wavelengths) and is called the electromagnetic spectrum. Details of the spectrum are given in table 1.1. Broadly speaking, the range may be divided into audio waves, radio waves (including television and radar waves), heat waves, light waves and the upper region of frequency which yields ultra-violet rays, X-rays and cosmic rays. (The word wave is replaced by ray in this region.)

Audio waves are those which when received and processed by suitable equipment give rise to sound waves that can be heard by the ear. Audio waves as such cannot be heard, because they are solely an electromagnetic disturbance; however, if alternating currents changing at frequencies within the audio range are fed to a loudspeaker, sound waves are generated which can be heard. Sound waves as distinct from electromagnetic audio waves will be considered in more detail in the next article.

Electromagnetic waves at audio frequencies are difficult to propagate and it is usual to employ higher frequency waves in the radio frequency band to carry audio intelligence, using a process called modulation. This process is described in article 1.6. The range of frequencies used as carriers, sometimes called *Hertzian* waves after their discoverer, extends from just above the audio range at about 20 kHz to just below visible light, the infra-red region beginning at around 750 GHz. Above the radio wave region there are radiant heat waves, light waves and the ray region. The visible light spectrum is quite small and it should be noted that, unlike electromagnetic audio waves, *electromagnetic waves at light frequencies* do produce the sensation of light directly on the human sensory organ, the eye.

As the frequency is further increased above the light region the waves become more penetrating and can cause physical changes in the human body. Ultra-violet radiation causes skin changes which in the mild form result in tanning of the skin and in the severe form can result in burning. The ability of X-rays to penetrate human tissue is well known and intense forms of radiation at these frequencies are used to destroy living matter. The effects of prolonged exposure to cosmic rays, from which earthbound humans are protected by the atmosphere, are still being investigated during lunar and other extra-terrestrial missions.

1.4. Sound Waves

The sensation of sound as recognised by the human brain is caused by the vibration of matter being passed to the appropriate nerve endings. The human ear consists of a diaphragm, a tightly stretched skin in the outer ear, which vibrates and passes this vibration via a series of small bones in the inner ear to the nerve endings. The vibration of the diaphragm is set up by movement of air or other particles surrounding the outer ear.

A medium of some kind is essential for the transmission of sound. This can easily be demonstrated by placing a ringing bell inside an inverted bell jar and evacuating the jar. As the air is removed the sound dies even though the bell is still observed to be ringing. The sound is restored as air is allowed to re-occupy the jar. The velocity of propagation of sound waves depends to some extent on the medium through which they are travelling; in air the velocity is approximately 330 metres per second (about 750 miles per hour).

As already stated, alternating current or electromagnetic waves at audio frequencies cannot be heard by the ear, but must be converted into sound waves by a suitable transducer (for example, a loudspeaker). A loudspeaker is a device in which an a.c. signal causes a coil to move a diaphragm in sympathy with the alternations of the signal. This vibration is then passed on via the air to the diaphragm of the ear. If a single-frequency signal (for example, a pure sine wave) is fed to a loudspeaker the sound vibration will be at the frequency of the signal, but the extent of the vibration, and thus the loudness, will be determined by the amplitude of the signal. The tone (or pitch) of the sound is determined by the frequency, the highest frequency producing the highest tone. The highest audible frequency varies from person to person and is determined by several things, particularly the age of the listener. The limit for the human ear rarely exceeds about 20 kHz, but certain animals, especially dogs, can exceed this limit.

Most sounds are not pure: that is, they are made up of a number of vibrations of various amplitudes and frequencies. Even if the sound is produced initially by an a.c. signal, the waveform will not be of sine-wave form. The alternating current waveforms corresponding to the 'o' and 'oo' sounds, for example, are illustrated in figure 1.9. Musical instruments depend upon these multi-frequency components for the variety of sounds that they produce and what is apparently the same note from two different instruments sounds slightly different, giving the sound the characteristic by which an expert can distinguish that particular instrument. The human voice also has this characteristic of many components, each one being determined by the length, and certain other features, of the vocal chords of the speaker. An electronic system designed to convey audio intelligence must be capable of allowing the passage of all component frequencies of the sound to be transmitted if faithful reproduction of the original is to be achieved. The telephone system is an example of a system which, for reasons of economy, is not designed to carry equally all component frequencies, and this is the reason why certain voices often appear modified when using that system. For the bulk of everyday speech sounds, of course, the telephone is quite adequate and intelligible conversation is usually possible. Faithful reproduction of music or high pitched voices, however, is restricted by the upper frequency limit of the system.

The three important characteristics of any sound, then, are *loudness, pitch* and *quality*. If the sound is produced by an a.c. signal, the loudness depends upon the amplitude of the largest component present, the pitch is determined by the highest frequency present, and the quality varies with the number of component frequencies present.

1.5. Complex Waveforms

Figure 1.9 shows the waveforms of alternating currents which, if supplied to a loudspeaker, would yield the sounds 'o' and 'oo'. As the resultant sound is composed of a number of component vibrations, so the a.c. signal which yields the sound can be considered to be made up of a number of component

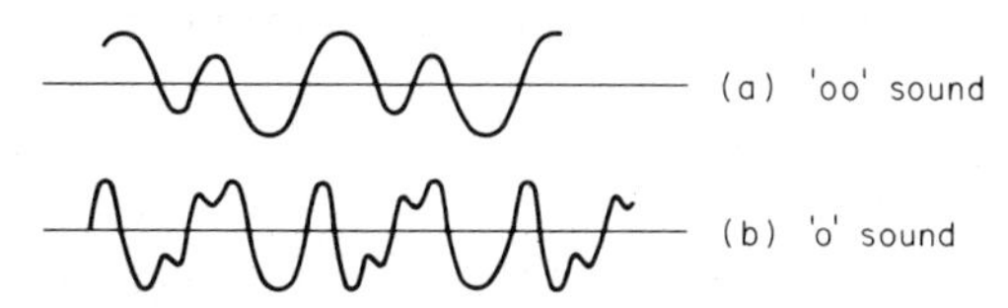

figure 1.9 a.c. waveforms corresponding to certain sounds

a.c. signals, each of which is of pure sine-wave form. Figure 1.10 is a further example of this and shows how an approximate square wave is made up of a number of component sine waves. In this example the frequency of each of the component sine waves is some simple multiple of the frequency of the largest component (called the fundamental frequency). The components are called *harmonics* of the fundamental frequency; a harmonic having a frequency equal to twice that of the fundamental is called a second harmonic; a component having a frequency equal to three times that of the fundamental is called the third harmonic; and so on. Figure 1.10 shows that the sum of a fundamental frequency and its odd harmonics (three times the fundamental frequency, five times the fundamental frequency, seven times the fundamental frequency etc.) is a square wave. Note that the amplitude of each harmonic is progressively reduced as the order (that is, the number) of the harmonic is increased. It can be shown similarly that the sum of a fundamental frequency and its even harmonics yields a triangularly shaped wave.

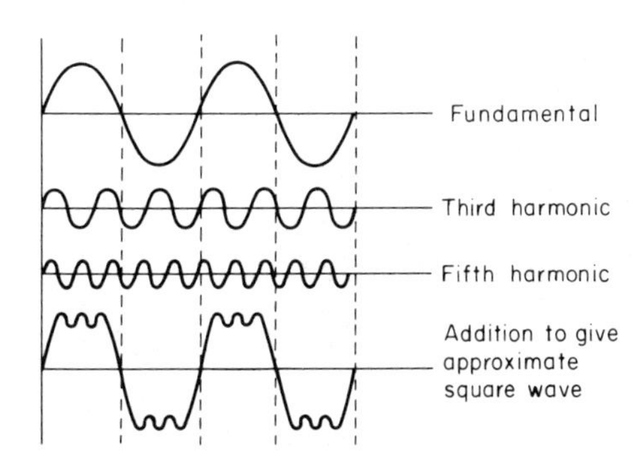

figure 1.10 Complex waveform–breakdown into harmonic components

It is useful to remember that any complex waveform can be broken down into components. The shape of any signal wave can be changed as desired by passing the signal through suitable circuitry which affects one or more of the components more than the others. Also, it is possible to obtain a resultant complex waveform by summing components (not necessarily harmonics). This process takes place during *modulation,* which is the process used to impress intelligence upon a carrier wave. This is discussed in the next article.

1.6. Modulation

As already stated, it is difficult to propagate electromagnetic waves at audio frequencies. This is largely due to the length of the aerial required and the high power that would be necessary to obtain efficient transmission. The normal method is to use electromagnetic waves at radio frequencies as carrier waves and to modulate (modify) these carriers with the intelligence signal to be transmitted. There are three kinds of modulation in common use: *amplitude modulation* (AM); *frequency modulation* (FM); and *pulse modulation.*

In amplitude modulation the intelligence signal is used to change the amplitude of the carrier wave. The carrier amplitude then varies as the amplitude of the signal to be carried. This is shown in figure 1.11a. The rate

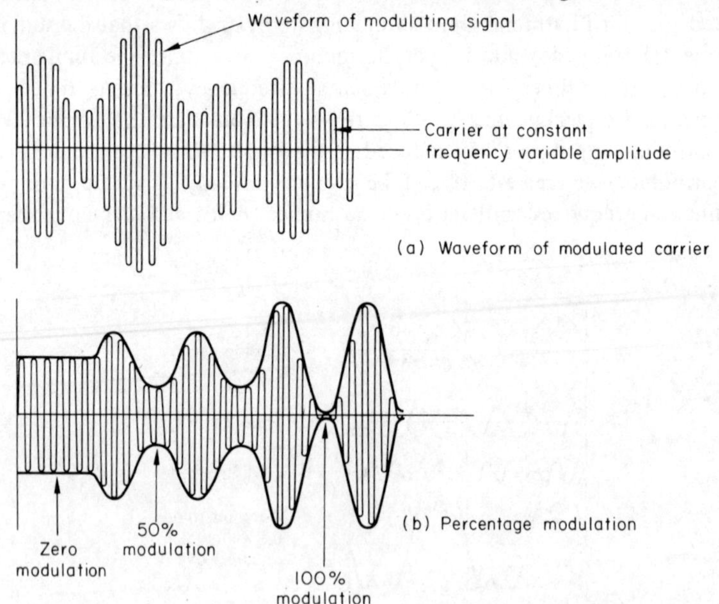

figure 1.11 Amplitude modulation

of change of the carrier amplitude is determined by the frequency of the modulating signal; the frequency of the modulated signal is held constant. The percentage of modulation of an amplitude modulated carrier is the amount by which the carrier amplitude changes relative to the amplitude of the unmodulated wave, expressed as a percentage. This is shown in figure 1.11b which shows a carrier modulated at different percentages. As can be seen, 50% modulation means that the carrier amplitude rises 50% above the unmodulated maximum value when the modulating signal rises to its maximum value in the 'positive' direction, and the carrier amplitude falls below the unmodulated maximum by the same amount when the modulating signal increases to its maximum value in the 'negative' direction. The maximum percentage modulation which can occur without distortion is 100%, when the carrier rises to twice the unmodulated maximum and falls to zero. Modulation above 100% distorts the carrier waveform from a pure sine wave and leads to the presence of undesirable harmonic frequencies.

Amplitude modulation gives a carrier of complex waveform and (as explained in article 1.5) such a waveform can be considered to be made up of several component waves at different frequencies. It can be shown that if an r.f. carrier of frequency f_c is amplitude modulated by a pure sine wave of frequency f_m, the resultant waveform contains three components at frequencies of $f_c - f_m$, f_c and $f_c + f_m$ respectively. For example, if a 1000 kHz carrier is amplitude modulated by a signal of frequency 4 kHz, the resultant modulated wave would contain three components at frequencies 996 kHz, 1000 kHz and 1004 kHz. If the modulating signal is itself complex (that is, it contains many frequencies), two side frequencies are generated for each component frequency of the modulating signal. Thus, the resultant modulated wave will contain a number of frequencies between the limits $f_c - f_{max}$ and $f_c + f_{max}$, where f_{max} is the highest frequency present in the modulating signal. The band of frequencies lying between the carrier frequency and the upper limit $f_c + f_{max}$ is called the *upper sideband,* and the band of frequencies between the carrier frequency and the lower limit $f_c - f_{max}$ is called the *lower sideband* (see figure 1.13). Circuits handling the modulated wave must be able to handle all component frequencies in both sidebands if faithful reproduction of the intelligence is to be obtained at the receiver. The *bandwidth* of such circuits (that is, the band of frequencies that can be handled) must then be equal to $(f_c + f_{max}) - (f_c - f_{max})$, that is $2f_{max}$ (or twice the highest modulating frequency).

A *frequency-modulated* carrier wave is shown in figure 1.12. In this type of modulation the carrier amplitude remains constant and the carrier frequency changes. The total change or *deviation* of the carrier frequency is determined by the amplitude of the modulating signal, and the rate of change

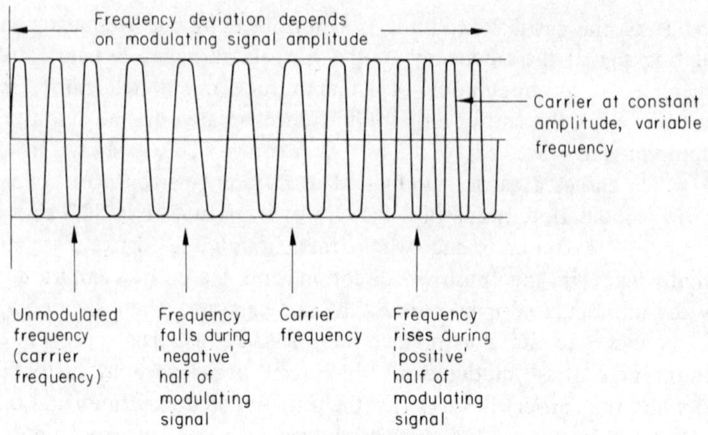

figure 1.12 **Frequency modulation**

of the carrier frequency is determined by the frequency of the modulating signal. As the modulating signal waveform increases in the 'positive' direction, the carrier frequency increases; and as the modulating signal waveform increases in the 'negative' direction, the carrier frequency falls. The range of carrier frequency from centre (unmodulated frequency) to highest, or from centre to lowest, is called the *maximum frequency deviation.*

As the carrier frequency is constantly changing the waveform cannot be a pure sine wave once the carrier is frequency modulated. The modulated waveform is thus complex and contains a number of component or side frequencies. With amplitude-modulation each frequency present in the modulating signal generates two frequencies in the modulated carrier, as already explained. With frequency-modulation, however, each frequency in the modulating signal generates a large number of frequencies in the modulated carrier. Using the notation above, a carrier of frequency f_c modulated by a single frequency signal f_m would contain side frequencies at $f_c + f_m$, $f_c - f_m$, $f_c + 2f_m$, $f_c - 2f_m$, $f_c + 3f_m$, $f_c - 3f_m$, $f_c + 4f_m$, $f_c - 4f_m$, and so on. The actual number is determined by the extent of the carrier wave distortion and thus depends upon the maximum frequency deviation. As this, in turn, depends upon the amplitude of the modulating signal, it is this factor which determines the number of side frequencies. The number generated for any given case can be calculated using a form of advanced mathematics known as Bessel functions, the theory of which is not within the scope of this book.

The highest or lowest side frequency generated per modulating frequency must not be confused with the highest or lowest frequency to which the carrier swings during modulation. The side frequencies are the frequencies of the pure sine waves which, when added together, give the complex waveform of the modulated carrier. The highest or lowest frequency to which the carrier swings is the frequency of alternation of the complex carrier waveform (a distorted sine wave) during a fractionally small interval of time. The side frequency range is usually far greater than that of the maximum frequency deviation.

If a complex waveform is used to frequency modulate a carrier, then sidebands of frequencies are produced as with amplitude modulation, but the limit of the upper sideband is not $f_c + f_{max}$ and the limit of the lower sideband is not $f_c - f_{max}$. The upper sidebands exist between f_c and $f_c + f_{max}$, then between $f_c + f_{max}$ and $f_c + 2f_{max}$, then between $f_c + 2f_{max}$ and $f_c + 3f_{max}$ and so on to a maximum determined by the amplitude of the largest component of the modulating signal. A similar situation exists in the lower sidebands (see figure 1.13). As can be seen, a series of sidebands exist, each having an upper or lower limit situated f_{max} hertz from the next. It should be noted that of all sidebands produced, only those closest to the carrier frequency have any significant effect; that is, the amplitudes of the side frequencies are considerably reduced as they move further away from the carrier frequency.

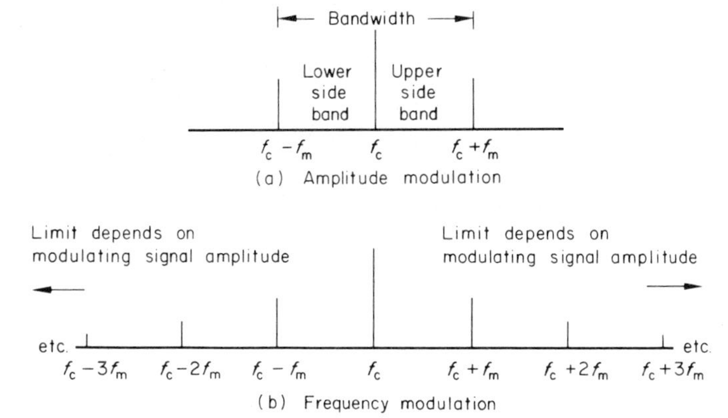

figure 1.13 Side bands: vertical lines indicate relative amplitudes (not drawn to scale)

In *pulse modulation* (illustrated in figure 1.14) the carrier is transmitted in pulses, rather than continuously as in the previous modulation systems considered. These pulses are then modified (modulated) in one of three ways: *pulse-amplitude modulation; pulse-width modulation*; and *pulse-position modulation.* In the first of these the intelligence to be transmitted is used to change the amplitude of the pulses, in the second to change the duration of the pulses, and in the third to change the position of the pulses relative to

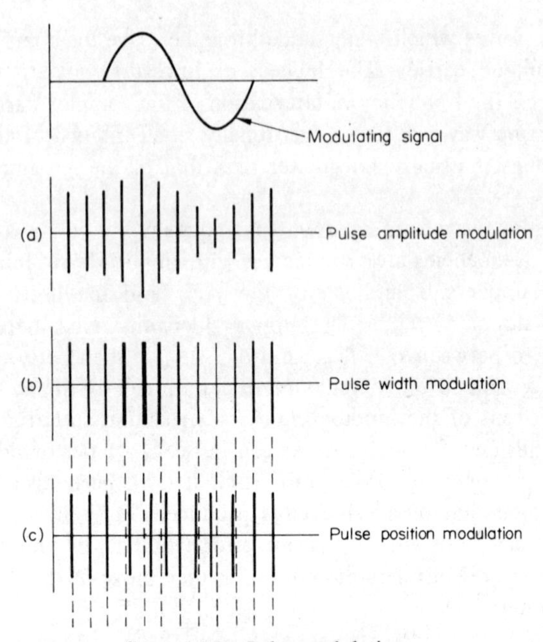

figure 1.14 Pulse modulation

their 'normal' unmodulated positions (that is, to arrive either before or after the 'normal' times of the unmodulated pulses as the modulating signal goes 'positive' or 'negative'). In all three types the extent of the modulation is governed by the amplitude of the modulating signal.

All types of modulation will be further discussed and compared as particular systems and specific circuits are examined in subsequent chapters.

2 System subunits

Any electronic system may be considered to be made up of a number of subunits, each one containing a combination of active and/or passive components (discussed in chapter 4) designed to process appropriately the electronic signal as it moves through the system. These different processes include generation, amplification, differentiation, integration, modulation, de-modulation and switching. Also all subunits containing active components require electrical energy and this is provided by a further subunit, the power supply. The power supply does not itself generate or process a signal, but wihout it no generation or processing using active components can take place.

Electronic subunits can be considered initially without detailed knowledge of their circuit arrangements. This chapter is concerned with a general discussion of the overall characteristics of subunits. Subsequent chapters will provide a more detailed account of individual circuits.

2.1. Power Supplies

Regardless of whether a.c. or d.c. signals are being transmitted through the system, most circuits containing active components (valves, transistors, diodes etc.) require a power unit that supplies direct current (that is, current flowing in one direction). An obvious method of providing d.c. is by using batteries and a number of systems, especially if they are to be portable (radios, record-players etc.), employ them. However, for larger fixed systems requiring rather more current, batteries are not convenient and d.c. power supplies derived from the main or other locally-generated electricity supply are used. Mains electricity in most parts of the United Kingdom comes in the form of alternating current and consequently provision must be made within the power supply circuit to change the a.c. to d.c. This process is called *rectification.* There are two kinds of rectification as illustrated in figure 2.1. In one, the sine waveform of the supply (figure 2.1a) is modified by removing

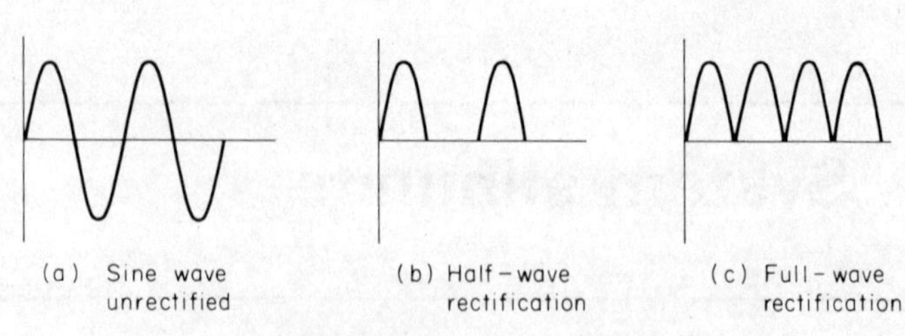

figure 2.1 Sine wave rectification

either the upper or lower half (as shown in figure 2.1b); this process is called *half-wave* rectification and is somewhat wasteful since no use is made of half the supply wave. In the other rectification process, one half of the wave is inverted as shown in figure 2.1c, the resultant current being a series of unidirectional pulses (that is, pulsating d.c.); this process is called *full-wave* rectification.

Both half-wave and full-wave rectification produce a direct voltage which is fluctuating. The waveform of such a voltage can be considered to be made up of a constant direct voltage and an alternating voltage. The direct-voltage component is the one required by the subunits fed by the supply, but the alternating component, or *ripple,* must be removed if the subunits are to operate correctly. This is done by means of a *filter* circuit which allows d.c. to pass virtually unaffected but blocks the flow of a.c. Figure 2.2 shows the

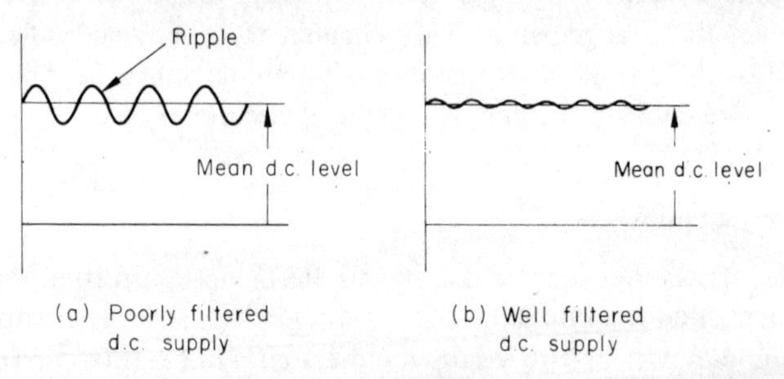

figure 2.2 Effect of filtration

waveforms of a poorly filtered output voltage and a well-filtered output voltage. The ideal output voltage would have no alternating component, but it is not possible, however, to completely remove the ripple. It should be noted that the ripple frequency is equal to the input-supply frequency for a half-wave rectifier and equal to twice the input-supply frequency for a full-wave rectifier.

The electrical circuits fed by a power supply are called the *load* on the supply. When a load is connected to a power supply the output voltage changes from the no-load value. An important property of any power supply is its *regulation characteristic* which is a graph showing how the output voltage changes as the load is increased. A poorly regulated supply gives an output voltage that is extremely dependent on load; such supplies are used only when the load is reasonably constant. If the load is likely to vary yet a fairly constant output voltage is required, a regulator circuit or *stabiliser* of some kind must be included between the supply and the load. Voltage regulation curves for a power supply with and without a stabiliser are shown in figure 2.3. Figure 2.4 shows a block diagram of a power supply showing the rectifier, filter and regulator blocks. Most regulators stabilise the supply not only against load variations but also against certain changes in the incoming a.c. supply.

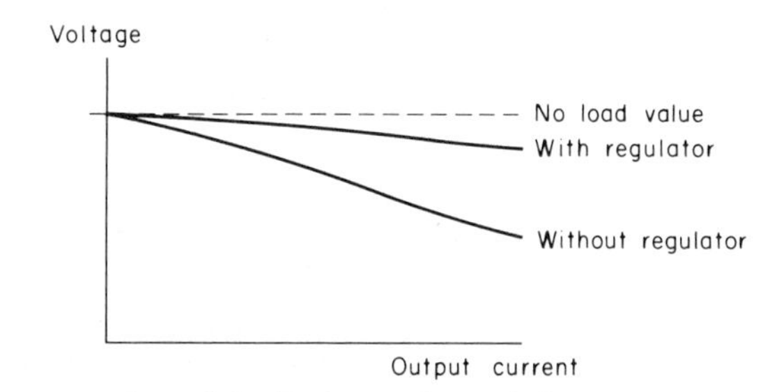

figure 2.3 Power supply regulation curves

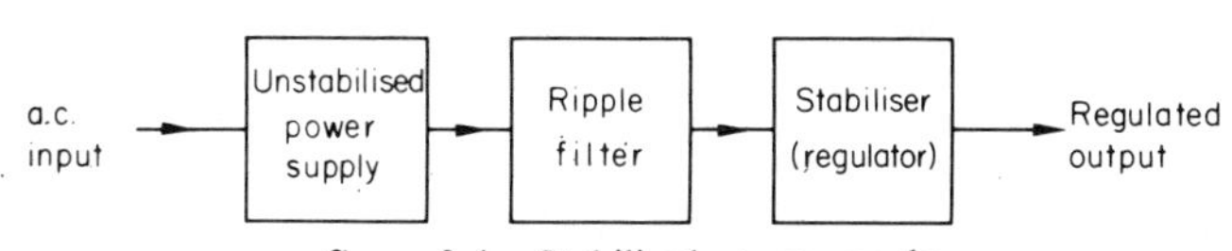

figure 2.4 Stabilised power supply

The system illustrated in figure 2.4 is a simple stabilised power supply. To obtain even better regulation (that is, a more constant output despite varying loads or varying inputs) a more sophisticated system of the kind shown in figure 2.5 may be used. In this type, two d.c. supplies are derived from one a.c. input supply. One d.c. supply is used to provide a constant reference voltage which is independent of load variation and which is proportional to the *desired* output. The other d.c. supply provides the *actual* output. The values of the actual output and the desired output are compared in a subunit called a *comparator,* the output of the comparator being used to control the regulator that determines the *actual* output. Should the actual output change, the difference between the new value and the desired value is measured by the comparator which then adjusts the regulator and thereby causes the

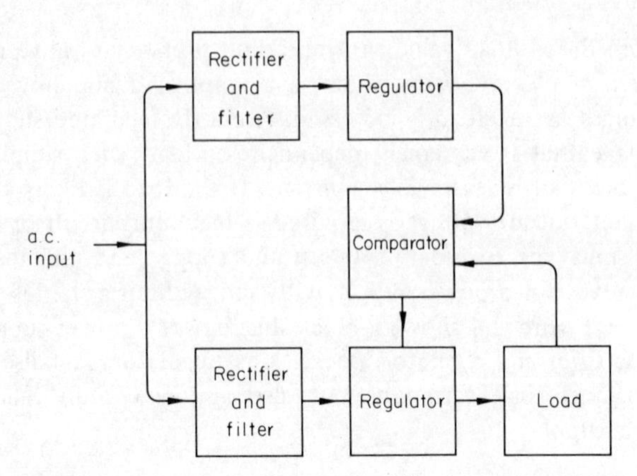

figure 2.5 A more sophisticated stabilised power supply

change in actual output to be reduced to as near zero as possible. The system can be used to provide either a constant voltage or a constant current supply as desired, the reference and output voltages in the latter case being proportional to the desired current and actual current respectively. This process of using the output to control itself is called *feedback.* Feedback of one form or another is used extensively in electronic systems.

The remaining subunit likely to be found in electronic power supplies is the *distributor* which is a circuit so arranged that various levels of voltage can be tapped off from the supply simultaneously. A distributor may be placed before or after the regulator, depending upon which supplies require to be stabilised and which do not.

2.2. Amplifiers

One of the most common subunits used for processing electronic signals is the *amplifier.* An amplifier strengthens the signal by increasing the amplitude of the voltage or current or by increasing the power available from the signal. Thus, there are *voltage amplifiers, current amplifiers* and *power amplifiers.* A power amplifier, for example, does not necessarily amplify voltage, but the voltage-current *product* (that is, the power) is greater at the output than at the input of such an amplifier.

As well as being defined in terms of the particular signal property being amplified, amplifiers are also sometimes described by the range of signal frequencies that they are capable of handling. These categories are d.c., a.f., r.f. and wideband. A *d.c. amplifier* handles signals that change at a very slow rate (not more than a few cycles per second); *a.f. amplifiers* handle the audio range of frequencies (up to 20 kHz); and *r.f. amplifiers* handle a narrow band of frequencies within the radio-frequency part of the electromagnetic spectrum. In contrast, a *wideband amplifier* is capable of handling signal frequencies from a few hertz to many millions of hertz; so that this kind of amplifier is used for handling square and other non-sine waves. As indicated in chapter 1, this kind of waveform contains high-frequency harmonics as well as the lower frequency fundamental.

There are seven properties of an amplifier which are of importance and which are used when comparing one circuit with another. These are the input impedance, output impedance, transfer characteristic, gain, frequency response, phase shift and feedback.

The word *impedance* means opposition to the flow of electric current. At zero frequency (that is, direct current) impedance is called *resistance.* The higher the value of impedance or resistance, the greater is the voltage developed across it for a particular current value. Similarly, for a particular value of applied voltage, the higher the value of impedance or resistance, the lower is the value of current that flows.

The *input impedance* of an amplifier or other subunit is the effective impedance between the input terminals as presented to the signal. Effective means that the impedance is not necessarily that of the component(s) seen to be connected across the input, but is a combination of that visible impedance with the impedance of the first active device (transistor, valve etc.) in the circuit. The active device impedance is influenced by a number of factors, including what feedback (if any) is present in the circuit. If internal feedback is present, the load connected to the amplifier output can considerably affect the effective input impedance. The magnitude of the input impedance is an important factor in the *matching* of one subunit to the next. This aspect is now further considered.

The *output impedance* is the *effective* impedance across the output terminals as seen when looking back into the output. As with input impedance, the output impedance is not necessarily that of the component or components connected across the output, but is a combination of that apparent impedance with the impedance of the last active device in the circuit as a whole. With certain active components, notably transistors, the value of the output impedance may be particularly affected by what is connected to the amplifier input.

Matching of an amplifier, or other subunit, to another means arranging the values of the output impedance of the one and the input impedance of the other so that there is as little as possible loss of signal during transfer. Consider figure 2.6 which shows a generator in series with two resistors R_o and R_i. (A resistor is a passive component having resistance: see chapter 4.) The generator and the resistor R_o represent the equivalent circuit of an

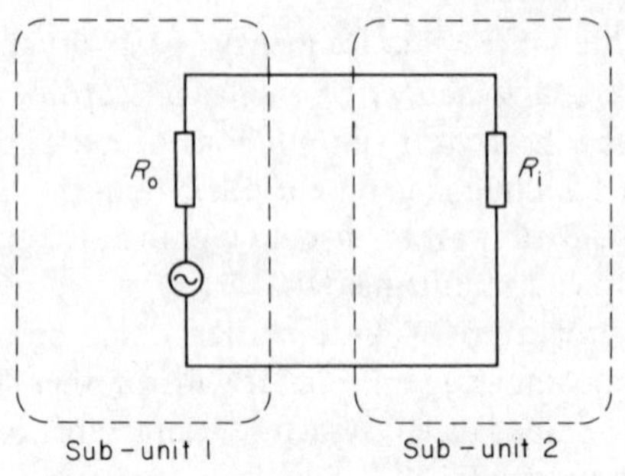

figure 2.6 **Input and output resistance**

amplifier and its output resistance, or the *effective* circuit as seen by the following subunit; the resistor R_i represents the input resistance of the next subunit. To obtain maximum voltage the input resistance should be high and the output resistance low, because the signal voltage to be transmitted through the system is that across the input resistance. Similarly, if the signal current is to be high the input resistance should be as low as possible; while if signal power is to be high the *maximum power transfer theorem* shows that the input and output resistance should be equal. As can be seen, then, the values of output and input impedances or resistances have a considerable influence on the effectiveness of the matching between one subunit and the next.

The *transfer characteristic* of an amplifier is a graph in which output voltage or current is plotted against input voltage or current (see figure 2.7a). The slope of the transfer characteristic is a measure of how much the amplifier amplifies (that is, of the *gain* of the amplifier): the steeper the slope, the larger is the amplitude of the output for a given input. The gain of an amplifier is determined by the type of active components used and by the pattern in which they are interconnected (this will subsequently be considered in greater detail). When the transfer characteristic is a straight line, the output waveform is a larger version of the input and the input waveshape is faithfully reproduced at the output. A curved transfer characteristic leads to a distorted output (as shown in figure 2.7b). A signal that is distorted when it leaves the amplifier must contain frequencies that the original input did not have, and the effectiveness of the transmission of intelligence is thereby reduced. In practical terms, if the system is a radio receiver, then the output sound will not be a true reproduction of the input; in a television receiver the picture at the screen will not exactly duplicate that seen by the camera.

As already indicated, the *gain* of an amplifier is a measure of its amplification. More precisely, it is the ratio of output signal amplitude to

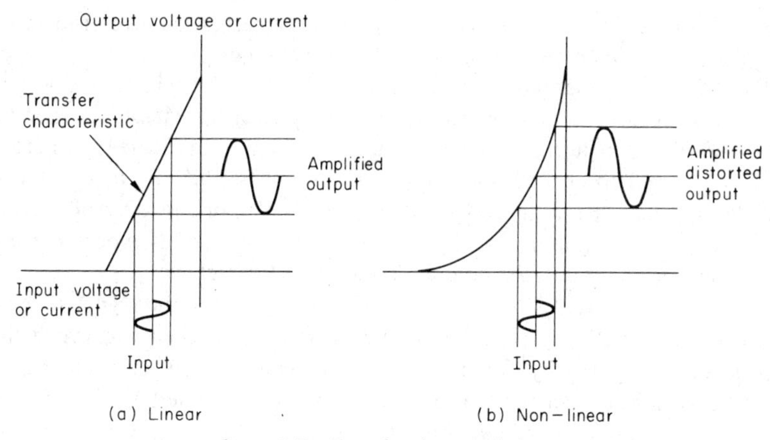

figure 2.7 Transfer characteristics

input signal amplitude (voltage or current) or, for a power amplifier, the ratio of output power to input power. Amplifier gain is determined by the individual amplification of the active components used and by the pattern in which they are interconnected, and it varies with frequency. A gain versus frequency graph (known as the *frequency response* curve) may be drawn for any particular amplifier and used to compare its performance with any other amplifier. Typical frequency response curves for audio and radio frequency amplifiers are shown in figure 2.8. The separation between the frequencies at which the gain falls to 0.707 of the maximum for a voltage or current amplifier, or to 0.5 of the maximum for a power amplifier, is called the *bandwidth.* An r.f. amplifier is described as a narrow band amplifier because the bandwidth is small; this is shown in figure 2.8b. Such a small bandwidth represents an amplifier that is very selective and this is precisely what is required in a subunit designed to select and amplify one carrier wave out of

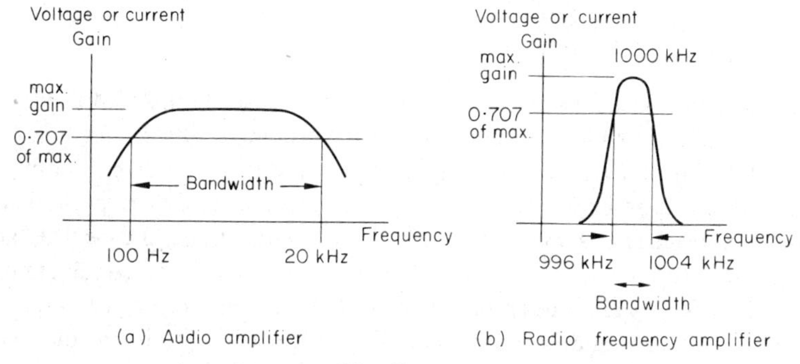

figure 2.8 Amplifier frequency response curves

many hundreds. The shape of frequency response curves can be considerably changed by the application of *signal feedback* techniques.

Phaseshift in an amplifier is the amount (if any) by which the output signal is delayed in time with respect to the input signal. Phaseshift may be expressed either in terms of time or, more commonly, in 'degrees', where 1 cycle is assumed to occupy 360 degrees. Thus, a phase shift of one quarter of a cycle (that is, where the output maximum occurs one quarter of a cycle after the input maximum) could be expressed as $T/4$ seconds, where T is the time taken for one cycle (the periodic time), or as 90 degrees.

Phaseshift is caused not only by the time taken for the signal to pass through the amplifier circuits but also by certain components in which the impedance is affected by frequency. This is further discussed in chapter 4, which covers both active and passive components. Whether or not the phaseshift of an amplifier is important is determined largely by the function of the signal being amplified. In switching systems, for example, in which certain switching actions must take place at a particular time, phaseshift may be of the utmost importance and steps may have to be taken to compensate for its effects. (See figure 2.9.)

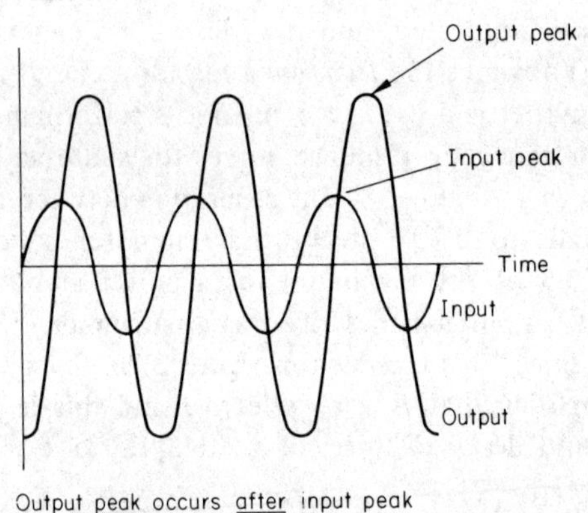

figure 2.9 Amplifier phase shift

The feedback in an amplifier is usually expressed in terms of the amount of output signal which is fed back to the input. Often, feedback is deliberately introduced into an amplifier, becuase it is found that amplifier performance can be improved in this way. In some active devices, particularly transistors, feedback is *inherent* (that is, it occurs *inside* the device) and special precautions must be taken to counteract it in those circumstances where its effects are undesirable. Feedback is an important technique in electronic systems and a more detailed account is given in the next four paragraphs.

Feedback is the process of taking either part or all of an output signal and feeding it back to the input. The feedback signal can be arranged to increase the input signal (the process is then called *positive feedback*) or to reduce the input signal (the process is then called *negative feedback*). Positive feedback, since it increases the input signal, produces a larger output signal and so the effective gain of the amplifier is increased. Similarly, negative feedback produces a reduced output signal and the effective gain is decreased. This is clarified in figure 2.10, which shows a voltage amplifier with and without feedback. Figure 2.10a shows an amplifier of gain 2 so that an input of 2 V produces an output of 4 V. In figure 2.10b part of the output is taken back to *oppose* part of the input (that is, the feedback signal is in *antiphase* with the input); the initial input of 2 V is now reduced to an effective 1 V giving an output of only 2 V. The effective gain is now 1, as an input of 2 V to the feedback amplifier gives an output of only 2 V. This is *negative* feedback as the effective gain is reduced.

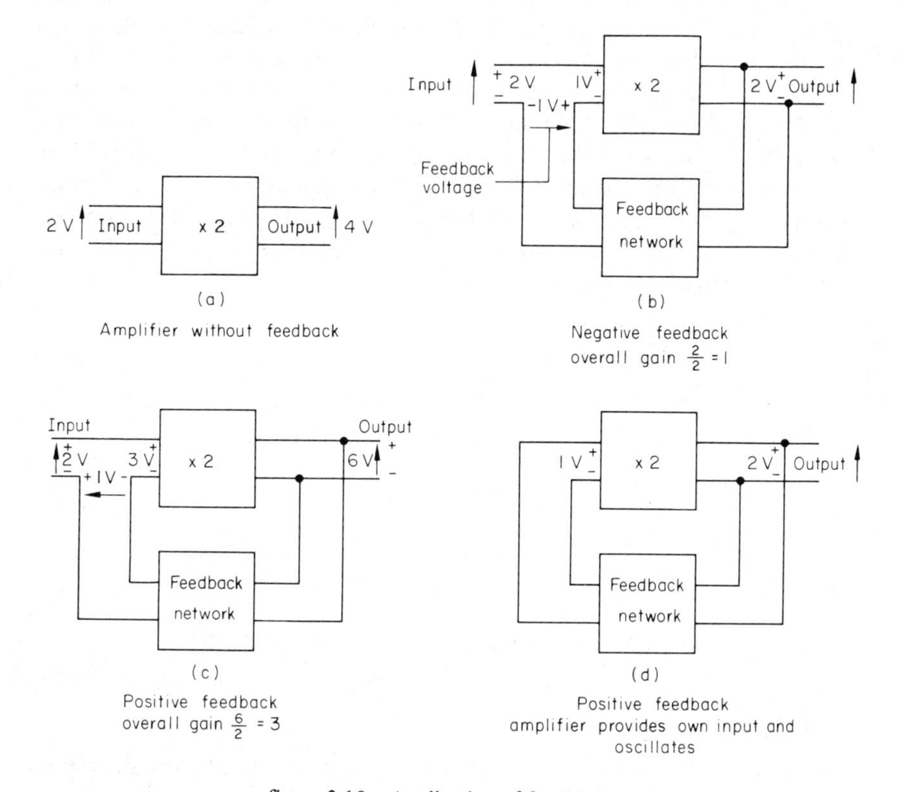

figure 2.10 Application of feedback

In figure 2.10c part of the output is fed back to assist the input (that is, the feedback signal is *in phase* with the input) so the amplifier receives a signal of 3 V. The output is thus 6 V because the amplifier gain is 2. The overall gain of the feedback amplifier, however, is increased to 3 as an input of 2 V (from the previous sub-unit) produces an output of 6 V. This feedback is *positive*.

A special case of positive feedback is illustrated in figure 2.10d. Here, part of the output is used to provide *all* the input and the amplifier generates its own signal. The effective gain under these conditions is infinitely high, because an output is generated without an input from a previous subunit. An amplifier connected in this manner is said to *oscillate* and the subunit is called an *oscillator.* Oscillators are widely used in the generation of electronic signals and are considered in the next article. The initial input to set an oscillator in operation is supplied by transient voltage or current surges that arise each time the oscillator is switched on.

The application of positive feedback to an amplifier merely to increase gain is not widely used because of the possibility of the subunit breaking into oscillation. Self-generation of a signal is obviously undesirable in a circuit that is designed to handle a signal already developed elsewhere. Sometimes positive feedback occurs internally within an amplifier and to avoid the possibility of oscillation negative feedback may be deliberately applied. This is not necessarily a bad thing, because although gain is reduced it can remain fairly constant over a much larger frequency range than it is able to do when feedback is not applied. Typical graphs of gain versus frequency with and without feedback are shown in figure 2.11.

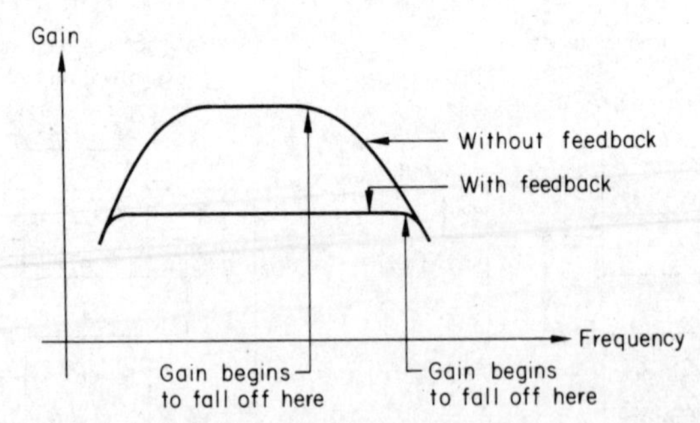

figure 2.11 Effect of negative feedback on frequency response

2.3. Oscillators

A subunit that generates a signal is called an *oscillator.* Oscillators are used to provide a.c. signals from frequencies just above zero to frequencies in the gigahertz region at the top end of the radio-frequency part of the electromagnetic spectrum. They may be described in one of two ways in terms of the type of signal waveform generated: *sinusoidal oscillators* and *relaxation oscillators.* A sinusoidal oscillator generates a signal having a sine waveform; a relaxation oscillator generates a signal that is usually of square waveform.

The sine wave oscillator comprises an amplifier that provides its own input by feedback through some kind of frequency controlling component such as a *tuned circuit* or a *resonant crystal* (as shown in figure 2.12a). These

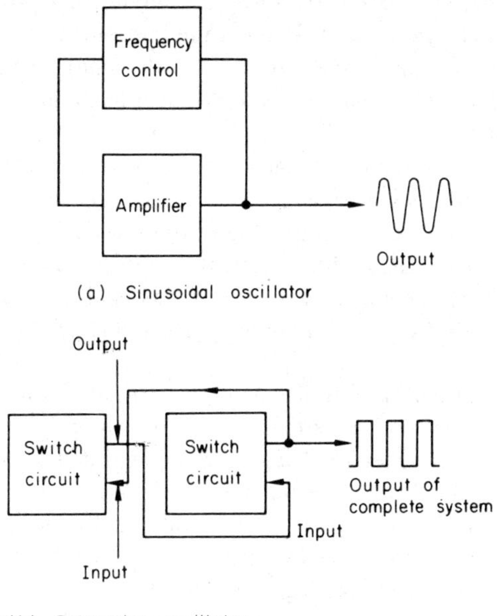

figure 2.12 Oscillators

components are considered in more detail in subsequent chapters. A relaxation oscillator comprises two switching circuits so interconnected that the output of each one in turn switches the other. The switching process is continuous so that the output waveform is that of a d.c. level being continually interrupted as shown in figure 2.12b.

Sinusoidal oscillators are widely used in radio communication for carrier generation and in many of the test instruments used with such systems. Relaxation oscillators are used as pulse generators in television and radar systems, in digital systems (computers and logic circuits) and in test instruments. Detailed circuits of both types of oscillator are presented in subsequent chapters.

2.4. Modulators, Demodulators and Detectors

Modulation was discussed in chapter 1. As explained there, it is a process by which intelligence to be conveyed is impressed upon a higher frequency carrier signal, the reason for the use of the process being that it is easier to transmit electromagnetic waves at frequencies higher than those of the intelligence. A subunit that carries out the modulating process is called a *modulator* and such subunits are found at the transmitting end of the system. A modulator has two inputs and one output: the inputs are the signal that is to be modulated and the signal that is to do the modulating; the output is the resultant modulated carrier.

Once received, the intelligence signal must be extracted from the carrier. A subunit to carry out this process is called a *demodulator* or *detector* and it has one input and one output, and the output carries the intelligence. The carrier wave is usually discarded by suitable filters, because once the signal is received the carrier has completed its task. A demodulator used to extract the intelligence from a frequency-modulated carrier is often given the name *discriminator*.

As was stated earlier, a radio frequency amplifier is designed to amplify a narrow band of frequencies centred about the carrier frequency that it is desired to receive. In order to change the required carrier (that is, to select another radio station) it must be possible to change the centre frequency to that of the new carrier. This process of moving the frequency response curve is called *tuning*. For weak incoming signals more than one r.f. amplifier may be necessary to strengthen the signal sufficiently before detection. As will subsequently be explained in more detail, it is not convenient to have a series of tunable amplifiers in one receiver. Consequently a process called *frequency conversion* is used, in which the carrier signal frequency (or carrier signal centre frequency for frequency modulation) is converted to a constant radio-frequency regardless of the value of the incoming signal frequency. This then means that subsequent r.f. amplifiers need not be tunable but can be designed to have a bandwidth centred about a single constant frequency. This constant frequency is called the *intermediate frequency* (i.f.). The r.f. amplifiers included to handle the i.f. are called i.f. amplifiers to distinguish them from the tunable r.f. types.

Frequency conversion, or *mixing,* uses a similar process to modulation, in that the incoming modulated r.f. is mixed with a second signal derived from an oscillator, to give an output waveform made up of a number of component frequencies, from which frequency-selective circuits are then used to extract the i.f. The i.f. signal is itself modulated in exactly the same way as the original r.f. but the carrier frequency is now different.

The name *converter* is usually given to a subunit that converts the r.f. to i.f. and contains its own oscillator. The name *mixer* is used for a subunit that requires a separate oscillator. (See figure 2.13.)

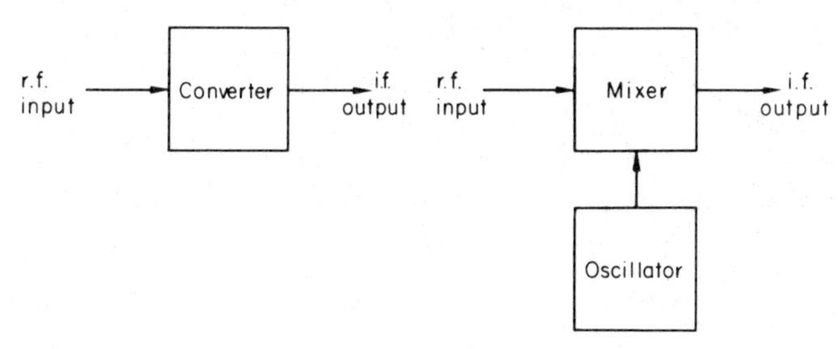

figure 2.13 Mixing and frequency conversion

2.5. Differentiators and Integrators

Differentiation is the name given to the mathematical process of determining the *rate of change* of a varying quantity. If a graph of the varying quantity is drawn the rate of change is given by the *slope* of the graph. For example, consider a voltage increasing with time at a constant rate as shown in figure 2.14a. This waveform is called a *ramp function* because of its shape. As the rate of change of a ramp function is constant, a graph of the rate of change looks like the one shown in figure 2.14b; that is, a straight line at a constant height above the time axis. If this straight line represented a voltage we would say that its waveform was that of a differentiated ramp function.

The opposite process to that of differentiation is called *integration.* Thus if the straight line waveform of figure 2.14b is integrated it gives the ramp function of figure 2.14a. With this process we are finding the waveform whose rate of change graph is the same shape as the waveform to be integrated. To further explain these two processes consider some typical signal waveforms as shown in the remainder of figure 2.14. The waveform in figure 2.14c is a square wave. The rate of change of such a wave is rapid and positive when the square wave rises, zero over the portion that the wave is constant, and rapid and negative when the square wave falls. If a square wave is differentiated the resultant waveform would then be that shown in figure 2.14d; that is, a series of alternate positive and negative pulses. Figure 2.14e shows the first part of a square wave: the sudden rise to a positive constant level. Such a waveshape is called a *step-function,* from its appearance. The rate of change of a ramp function is constant so that if a step-function is integrated the resultant wave is a ramp function as shown in figure 2.14f. Finally, a sine wave is shown in figure 2.14g. The rate of change of a sine wave is a positive maximum as it goes from negative to positive through zero, is zero when the wave reaches a maximum and is a negative maximum as the sine wave goes from positive to negative through zero. The

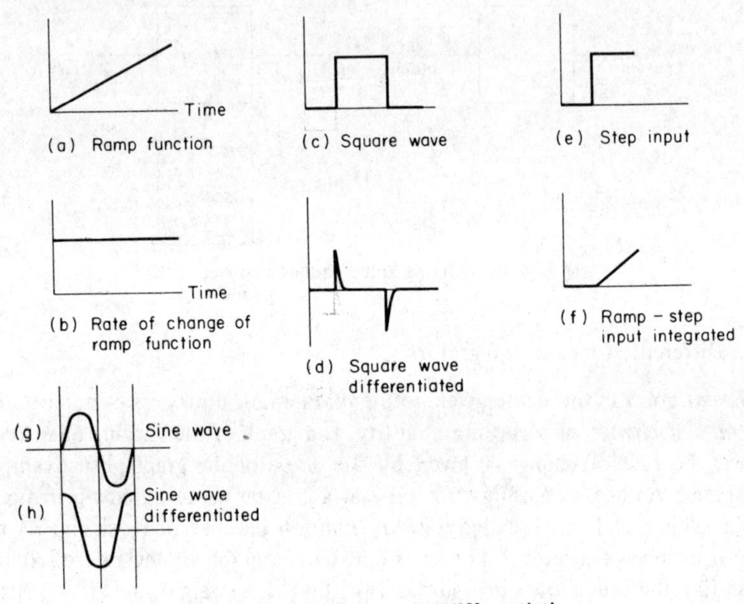

figure 2.14 Integration and differentiation

graph of the rate of change is, in fact, itself a sine wave but displaced a quarter of a cycle to the left (that is, a sine wave shifted *in phase* by the time taken for a quarter of a cycle). If a sine wave is differentiated the resultant waveshape is thus another sine wave but displaced in phase to the left. Such a phase displacement is called a *lead*; that is, the differentiated wave *leads* the original wave by a quarter-cycle (see figure 2.14h). Conversely, an integrated sine wave will be phaseshifted a quarter-cycle to the *right*; that is, will *lag* the original wave.

Integrating and differentiating circuits, called integrators and differentiators respectively, may contain all passive components or a combination of active and passive components. Under certain conditions certain kinds of amplifier may be used to carry out these processes. As with the other blocks, detailed circuits are considered later.

2.6. Limiters, Clippers and Clamps

Sometimes in certain electronic systems it is necessary to limit the signal amplitude to a particular level. A subunit that does this is called a *limiter.* One example of the use of a limiter is in an *FM* receiving system. As was described in chapter 1, frequency modulation of a signal means using the intelligence to be transmitted to modify the frequency of the carrier signal. The detector circuit (discriminator) used to extract the intelligence from the carrier as it proceeds through the receiver may well be sensitive to amplitude changes as well as frequency changes, in which case any amplitude change might be taken as part of the intelligence signal. This would produce distortion of the intelligence signal. An amplitude limiter inserted before the discriminator removes the problem.

Clipping is a similar process to limiting and often the two words are used to mean the same thing. Clipping consists of removing a section or 'slice' of a waveform as shown in figure 2.15. One use of the process is to obtain an approximate square wave from an original sine wave by taking the 'slice' shown in the figure and straightening the rise and fall parts of the signal. If there is a difference in the two processes, clipping and limiting, it is probably that a limiter retains the original waveform whilst controlling its amplitude; whereas a clipper, which also controls amplitude, does not necessarily retain the original waveform.

Clamping a signal either sets the d.c. level about which the signal waveform alternates, or prevents a signal from rising above a particular level. In the latter case clamping is similar to limiting or clipping, but only on one half of the signal waveform.

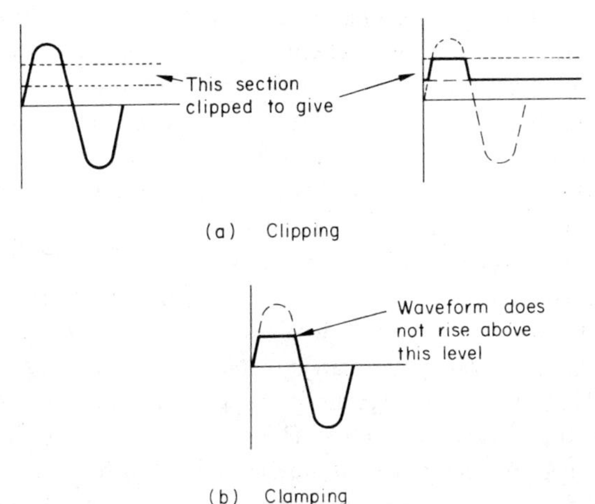

figure 2.15 Limiters, clippers and clamps

2.7. Logic Gates

An electronic *gate* is a subunit that allows a signal to pass under certain conditions. If these conditions are not satisfied the gate remains closed and the signal cannot proceed. The dictionary defines the word 'logic' as a 'connection or outcome of events' and it is in this sense that the word is applied to electronic gates; the connection of events being that if certain conditions are satisfied (that is, if certain events take place) then a further event (that of the signal proceeding) follows logically. The field of logic in electronic systems is very broad and logic theory is being applied more and more to all kinds of control, measurement and calculating systems. Logic theory is, of course, the basis of what is probabily the best known electronic system, the *digital computer.*

Logic circuitry in general comes under a branch of electronics known as *digital applications* as opposed to *linear* or *analogue applications.* Digital systems use signals which are of pulse form, that is, a series of short-duration square waves; analogue systems use signals of, or derived from, the familiar sine waveform. Amplifiers used in digital systems must be *wideband* amplifiers because of the many harmonics that make up a square wave.

Logic signals have a unique characteristic in that the signal is a d.c. signal that is always at one of only two levels. These levels are described as 'high' and 'low', or referred to numerically as '1' and '0'. If the low level (that is, the less positive level) is denoted by 0 and the high level by 1, *positive logic* is said to be in use. If 1 and 0 denote the low and high levels respectively, *negative logic* is said to be in use. The main advantage of using only two levels is that they are easily distinguished from one another (particularly if, as is usual, they are generated by switching a circuit either 'on' or 'off'). There can be little confusion in deciding whether or not a circuit is conducting. If several levels are used, however, confusion can arise due to the *drift* of a d.c. signal when passing through a system. Severe drift could cause a signal to change from one level to the next if several levels close to one another were used. It is, however, unlikely that drift would change a level corresponding to the 'on' condition to that corresponding to the 'off' condition.

There are *three* types of logical function performed by electronic gates: these are known as the AND, OR and NOT functions. Gates to carry out these functions are called AND gates, OR gates and inverters (or negaters) respectively. In addition, gates carrying out the NOT-AND and NOT-OR functions are called NAND gates and NOR gates respectively. All gates have one output and one or more inputs. The symbols for the different gates are shown in figure 2.16. The functions are defined as follows:

The AND function is performed when a gate prevents the output signal attaining logic level 1 until *all* the inputs are at logic level 1. The OR function is performed when a gate allows the output to attain logic level 1 if at least *one* input is at logic level 1. The NOT function is performed if the gate output is at the opposite level compared to the input (that is, if the output is 1 when the input is 0, or vice versa).

To clarify these definitions consider a two input AND gate. The possible combinations of inputs are as follows:

Input 1	Input 2	Output
0	0	0
0	1	0
1	0	0
1	1	1

This type of table showing input and output levels is called a TRUTH TABLE. The truth table shows that the output is 1 only when *both* inputs are 1. (See figure 2.16.) The truth table for a two input OR gate is as follows:

Input 1	Input 2	Output
0	0	0
0	1	1
1	0	1
1	1	1

Here the output is 1 when one (or more) input is 1 (see figure 2.16). The last line is the same as that of the two-input AND gate truth table above: the OR function includes the AND function. An OR gate that does *not* include the AND function is called an *exclusive* OR gate. The truth table for such a gate having two inputs is as follows:

Input 1	Input 2	Output
0	0	0
0	1	1
1	0	1
1	1	0

The symbol for an exclusive OR gate is shown in figure 2.16.
The truth table for an inverter is as follows:

Input	Output
0	1
1	0

A gate which carries out the AND function and then the NOT function is called a NAND gate. The truth table for two inputs is as follows:

Input 1	Input 2	AND	NAND Output
0	0	0	1
0	1	0	1
1	0	0	1
1	1	1	0

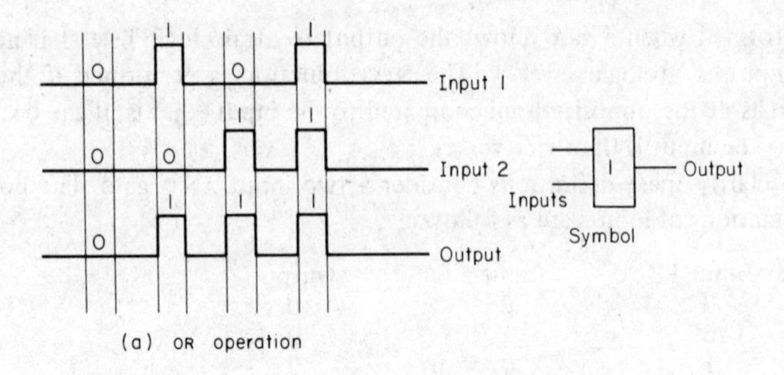

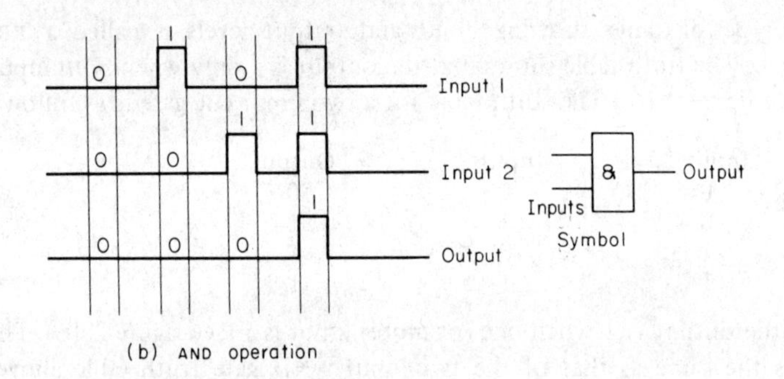

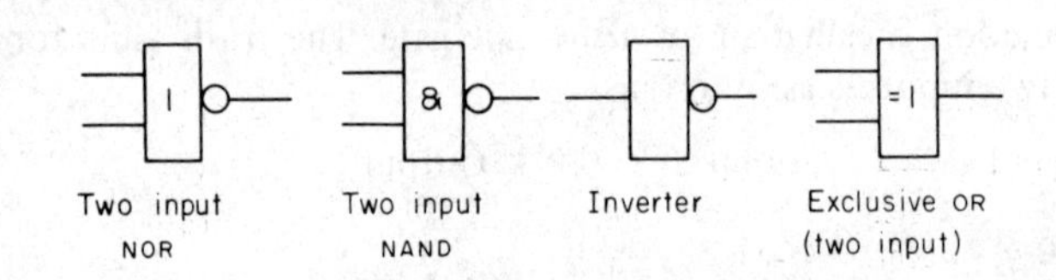

figure 2.16 Logic gate operation and symbols. Note that in parts (a) and (b) the most positive level is taken as logic level 1 (that is positive logic is used)

The output is the AND function *negated* (or *inverted*). Another name for the inversion process is *complementation,* so we can also say that the output is the *complemented* AND function.

A gate that performs the OR function and complements the result is called a NOR gate. The truth table for a two-input NOR gate is as follows.

Input 1	Input 2	OR	NOR Output
0	0	0	1
0	1	1	0
1	0	1	0
1	1	1	0

Changing the type of logic (that is, positive to negative, or vice versa) for a particular gate changes the function of that gate. Consider a gate that has the following truth table.

Input 1	Input 2	Output
Low	Low	High
Low	High	High
High	Low	High
High	High	Low

This means that if the signal level at inputs 1 and 2 is low, the output signal level is high and remains high unless *both* input signal levels are high. Positive logic denotes high by 1 and low by 0, so the truth table becomes as follows.

Input 1	Input 2	Output
0	0	1
0	1	1
1	0	1
1	1	0

which is the truth table of a two-input NAND gate. Negative logic denotes high by 0 and low by 1, so the truth table becomes as follows.

Input 1	Input 2	Output
1	1	0
1	0	0
0	1	0
0	0	1

which is the truth table of a two-input NOR gate. It can be seen therefore that the same gate is capable of a dual function, and this is the reason why most manufacturers describe their gates as NAND/NOR.

NAND/NOR gates may be used in numerous ways to make up complete systems capable of counting, measuring and computing. When setting up a complete system the truth table for the overall system is drawn up; this then gives the required output signal or signals for the various combinations of inputs. From the truth table equations are obtained using a form of mathematics known as *Boolean Algebra.* These equations are then simplified and the result gives the numbers and types of gate required and indicates how they must be interconnected to perform the desired overall function. Other subunits used in conjunction with the basic gates include flip-flops and multivibrators.

2.8. Flip-flops, Multivibrators and Memories

A *flip-flop,* or *bistable,* is a subunit having one (or more) outputs capable of holding a signal in either one of two possible states: the output can be at level 1 or 0 and hold this level indefinitely. On the receipt of an input trigger pulse, the output changes state to the complement of the previous state (that is, from 1 to 0, or vice versa) as in figure 2.17a. A flip-flop is thus capable of *storing* a signal level and can thereby act as a *memory*.

A *monostable* subunit has only one stable state. An input pulse will change the output to the complement but it will not remain in that state; after a certain time the output returns to the state that existed before receipt of the input pulse. The output waveform is shown in figure 2.17b.

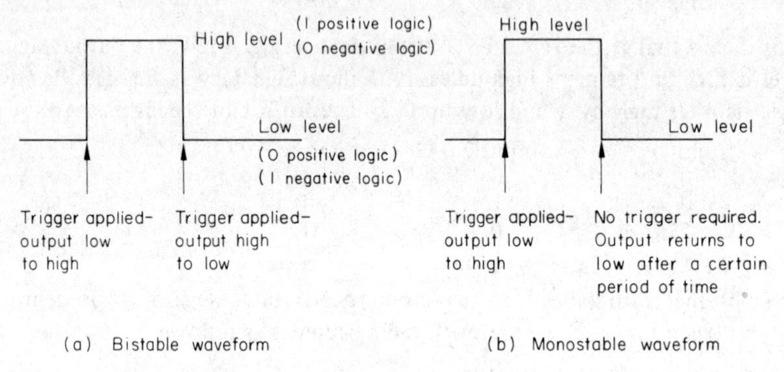

figure 2.17 Switching circuit waveforms

An *astable multivibrator* or *pulse generator* is a form of *relaxation oscillator* (already considered in article 2.3). This kind of subunit is used to generate the input signals used to close and open the various gates at the required time and in the required order, and it produces a pulse waveform of the type shown in figure 2.12b.

3 System block diagrams

Chapter 1 described the various kinds of electronic signal and in chapter 2 a number of the more commonly used subunits for signal processing were considered. This chapter is concerned with joining together the various subunits to form complete systems. Descriptions of actual circuitry follow in subsequent chapters.

3.1. Radio Systems

Figure 3.1 shows a block diagram of a simple transmitting system for the electromagnetic propagation of audio signals. It consists of six subunits, one of which is the power supply providing direct current to all remaining subunits. The master oscillator generates the signal carrier frequency which, after amplification in a radio frequency amplifier, is transferred to the modulator. The audio frequency intelligence signal is derived from a

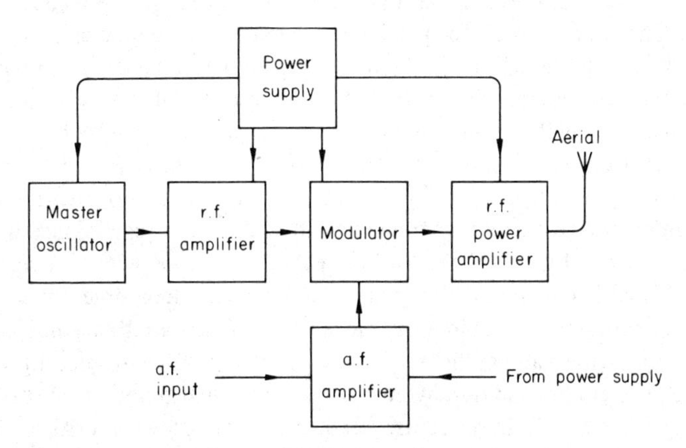

figure 3.1 Simple transmitter system

transducer (a microphone, record player pick-up, etc.), amplified in an audio amplifier and then transferred to the modulator. The a.f. signal is impressed upon the r.f. carrier using one of the available forms of modulation (AM, FM etc.) and the modulated carrier is then again amplified to provide a high power output to the aerial system. Figure 3.2 shows a similar system for AM transmission, and this will be seen to include a *buffer amplifier* between

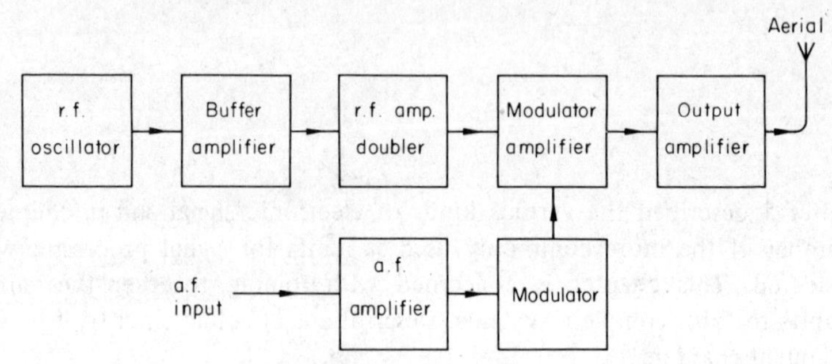

figure 3.2 AM transmitter

oscillator and r.f. amplifier. This subunit helps to reduce frequency drift of the carrier, an essential requirement in an AM system. The r.f. amplifier shown here is indicated as a *frequency doubler.* This subunit gives an output frequency equal to twice the input frequency. The use of such a subunit allows the use of frequency stability control devices such as piezoelectric crystals (see chapter 7) at a much lower frequency than that used for propagation.

The block diagram of an FM transmitter using a *reactance modulator* is shown in figure 3.3. This system also employs automatic frequency control (AFC). The master oscillator produces the carrier frequency as before. This output is then progressively doubled and tripled before being fed to the aerial system via a power amplifier. The master oscillator frequency is controlled by a reactance modulator, which puts a variable capacitance across the oscillator tuned circuit (see chapter 7). The value of this capacitance is in turn controlled by the audio amplifier that provides the audio intelligence. As stated in chapter 2, the carrier frequency then changes in accordance with the variation in amplitude of the audio signal. In the absence of a signal the carrier should return to the centre frequency. Sometimes this centre frequency drifts and this in turn moves the entire variation (that is, both sidebands) further along the frequency spectrum. This would, of course, create a problem in reception by an FM receiver tuned to receive an FM signal swinging about the given centre frequency. To prevent drift in centre frequency an AFC subsystem consisting of a second oscillator, mixer,

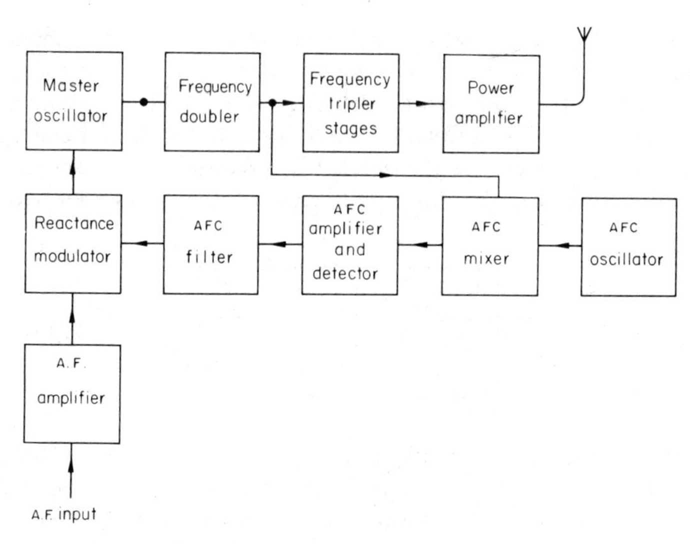

figure 3.3 FM transmitter with AFC

amplifier, detector (discriminator) and filter is employed. A signal for this subsystem is derived from an appropriate point in the main system (the output of the doubler, as shown in figure 3.3) and mixed with the fixed output of the AFC oscillator. (This oscillator is usually crystal controlled.) The mixer output is then amplified and fed to a detector that is tuned to the frequency which should be present when the correct carrier centre frequency is mixed with the AFC oscillator output. The mixer output is then fed via a filter to the modulator. If the carrier centre frequency changes, the mixer output changes and the detector produces an output voltage that controls the modulator, thereby returning the master oscillator to the correct frequency. When the carrier centre frequency is correct the discriminator produces zero output and no control voltage is applied to the modulator. It may not be immediately apparent how this system is able to differentiate between normal change in carrier frequency due to the audio signal and drift in carrier frequency due to the centre frequency shifting. Fortunately, these can be distinguished, since frequency drift is a slower process than normal modulation change. This is the purpose of the filter between discriminator and modulator, for it allows slow changes to take effect but does not respond to fast changes. Thus, the AFC system does not cancel the normal modulation but controls the master oscillator only when the carrier frequency drifts. A similar system may be employed in an FM receiver to ensure that the receiver stays tuned to the correct centre frequency.

A simple tuned radio frequency (TRF) audio frequency receiver is shown in figure 3.4. The signal is selected by tuning the radio frequency amplifier to give maximum gain at the desired frequency. The amplified carrier is then passed to a detector/filter circuit which extracts the audio signal and removes the r.f. component of the carrier. The audio signal is then further amplified to a level sufficient to drive a loudspeaker or other output transducer. As is explained in later chapters, this simple receiver suffers from poor *selectivity*

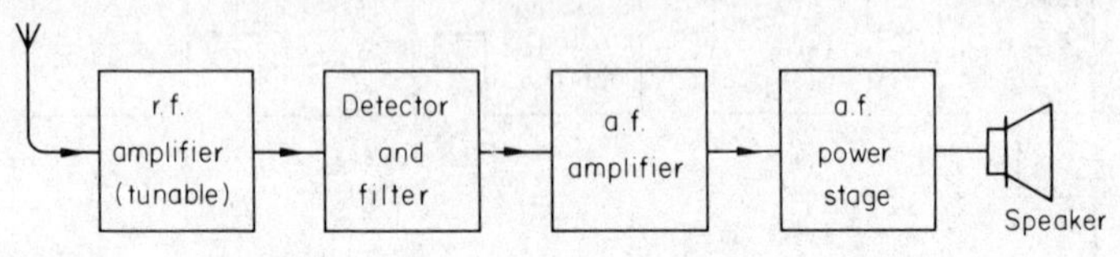

figure 3.4 TRF receiver

(that is, it is difficult to select a single carrier frequency free from interference by others) and *sensitivity* (that is, weak signals may be insufficiently amplified). The reason for this lies mainly in the fact that in a TRF system *all* r.f. stages must be tunable. In a *supersonic heterodyne* receiver (or *superhet* for short) only the input r.f. stage, local oscillator and mixer need be continuously tunable. Such a system for AM transmissions is shown in figure 3.5.

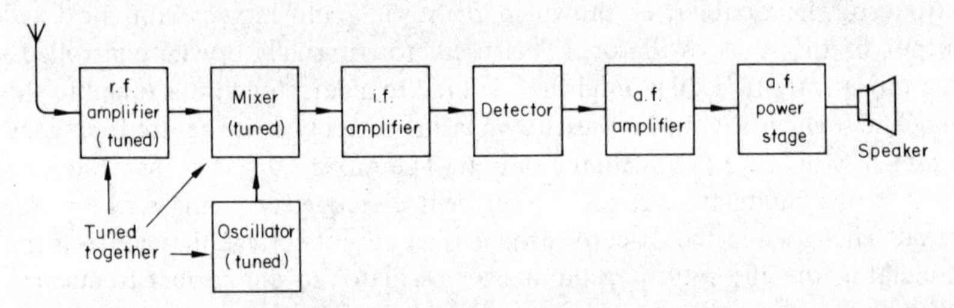

figure 3.5 AM Superheterodyne receiver

In the superhet system, the first r.f. stage is tuned to the desired carrier frequency, and a separate local oscillator stage is adjusted simultaneously to provide an output at a frequency that exceeds the frequency of the desired signal by a fixed amount called the *intermediate frequency*. Both the input r.f. and the local oscillator output are fed to a mixer stage which produces an output at a frequency that is equal to the *difference* between them; which is, of course, the intermediate frequency (i.f.). The i.f. modulation is exactly the same as that of the original r.f. input. The i.f. signal is now passed through a series of amplifiers set to give maximum gain at the intermediate frequency. These amplifiers are pretuned to the i.f. frequency, as whatever the input frequency from the aerial, the local oscillator is simultaneously adjusted to give an i.f. that is always the same. Cascading the i.f. amplifiers in the manner shown gives a very selective high gain system. From the i.f. amplifiers the signal is detected and filtered, and the a.f. is amplified in the same way as in the TRF system.

The use of the superhet principle is not confined to AM transmissions. A block diagram for a superhet FM receiver is shown in figure 3.6. The

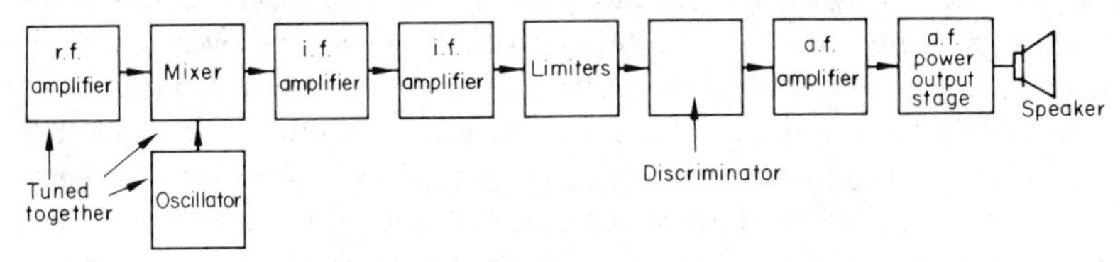

figure 3.6 FM Superheterodyne receiver

arrangement is similar to figure 3.5 with an additional *limiter* stage. This stage ensures that the signal remains at constant amplitude, for certain types of FM detector (discriminator) will respond to amplitude changes as well as frequency changes and this is, of course, undesirable. If AFC is incorporated in the system the discriminator output is taken to a reactance modulator or similar circuit, which is then used to provide additional control over the local oscillator. Thus, if the local oscillator drifts, the (relatively slow) change in the intermediate centre frequency is detected by the discriminator, and the local oscillator is returned to the correct setting for the desired input (that is, to a frequency that exceeds the input centre frequency by the value of the i.f.).

3.2. Television Systems

Television system signals are of necessity more complex than radio system signals. In a monochrome (black and white) television receiver, the electron beam in the picture tube—a form of cathode ray tube (see chapter 4)—moves rapidly from left to right across the screen, returns to the left-hand side at a point slightly lower down, then repeats the movement across to the right-hand side. In this way the beam creates a series of lines on the screen (405 or 625 in the UK). As the beam 'scans' the screen in this way the illuminating effect of the beam is increased or decreased as required (by the picture signal received from the television camera) in accordance with the light distribution on the scene that the camera is simultaneously viewing. The spot of light on the screen thus appears brighter or darker as the beam moves spot by spot over the full area of the screen, and the picture that is built up therefore duplicates that seen by the camera. The camera is basically similar, in that a battery of light sensitive 'cells' pointing at the scene to be viewed is scanned by a beam of electrons and the electric charge distribution on these 'cells', which is determined by the light distribution on the scene, is monitored by the beam to provide the picture signal. It is essential that the camera beam and the receiver beam scan in exactly the same way and are in the same relative position at the same time, otherwise the top left-hand corner of the view seen by the camera, for example, might appear at the bottom right-hand corner—or indeed anywhere—on the receiver screen. To achieve this, *synchronising (sync.) signals* are sent from the transmitter to the receiver so that the varying voltage used to drive the spot horizontally (the *line timebase* voltage) and the varying voltage used to drive the spot vertically (the *field timebase* voltage) are at the same relative values at the same time. (Note: the term 'field' has only recently begun to displace the older term 'frame'.) The composite *video signal* is thus composed of field and line synchronising signals and the picture signal (see figure 3.7a). The sound associated with the picture is transmitted separately by a conventional FM system at a different carrier frequency to that of the video signal.

The simplified block diagrams of a television transmitter and receiver for monochrome signals are shown in figures 3.7b and 3.8 respectively. In the

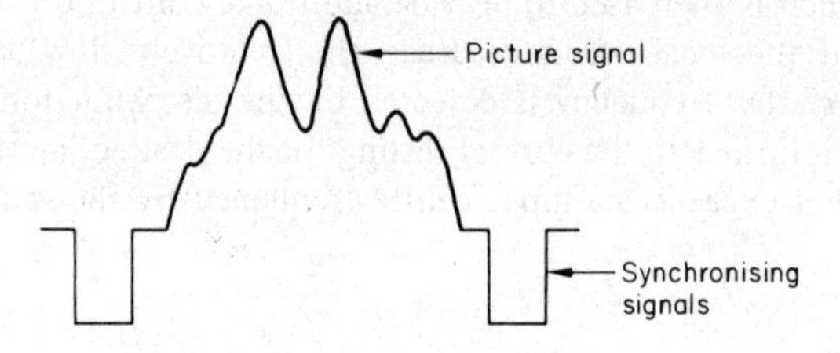

(a) Monochrome TV signal showing line sync. pulses

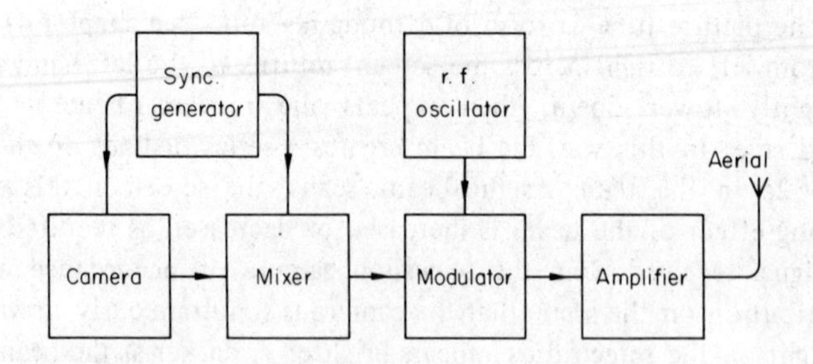

(b) Simplified monochrome TV transmitter

figure 3.7 Monochrome TV transmission

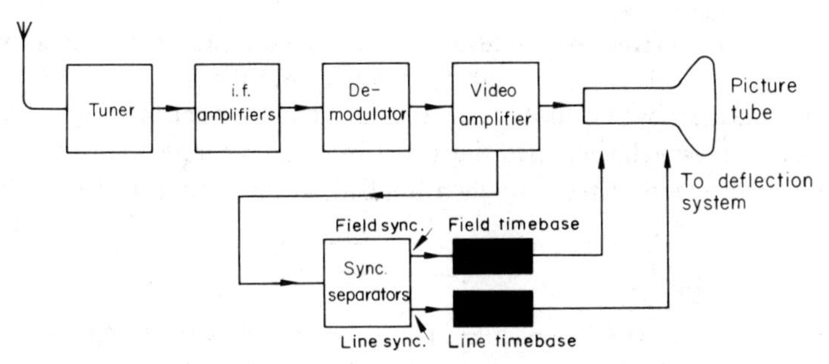

figure 3.8 Monochrome TV receiver simplified block diagram

transmitter the synchronising signals are fed from an appropriate generator to the camera and to a mixer unit. In this unit the picture signal determined by the light distribution seen by the camera is mixed with the synchronising signals to give the composite video signal. This signal then modulates the r.f. carrier and is transmitted in the same way as a radio signal. The receiver selects the required signal, detects (demodulates) it to produce the video signal, and this is then split up into the three parts, picture signal, field synchronising signal and line synchronising signal. The picture signal is fed to the electron gun of the picture tube (chapter 4), while the field and line sync. signals are fed to the tube vertical and horizontal deflecting systems respectively.

A colour television system uses a similar principle to the monochrome, in that the picture and sync. components make up the composite signal; but, as might be expected, the system is more complex, for three colour signals are involved instead of a single monochrome signal. The colours are red, blue and green, and from these a picture of all colours normally seen can be constructed. The picture tube commonly used has a complex screen composed of millions of dots of suitable material which glow red, blue or green when struck by an electron beam. The electron gun in these tubes produces three beams, one per colour. By using a device called a 'shadowmask', it is arranged that each beam strikes only its own series of dots (that is, the beam controlled by the 'red signal' strikes only the dots which glow red, and so on).

The composite signal in a colour TV system can be considered to be made up of four parts: the *luminance* component; the *colour component*; the *picture synchronising component* (made up in turn of line and field signals); and the *colour synchronising signal component.*

A simplified block diagram of a colour receiver is shown in figure 3.9. The luminance signal controls the brightness of each colour displayed on the screen and after amplification (controlled by 'brightness' and 'contrast' controls at the receiver) is passed to the picture tube cathodes. The colour or *chrominance* signal is amplified (controlled by the receiver 'saturation' control) and then passed first to a colour demodulator and subsequently to a colour decoder which separates the signal into the separate colour signals for red, blue and green. These are then applied to the appropriate grids in the

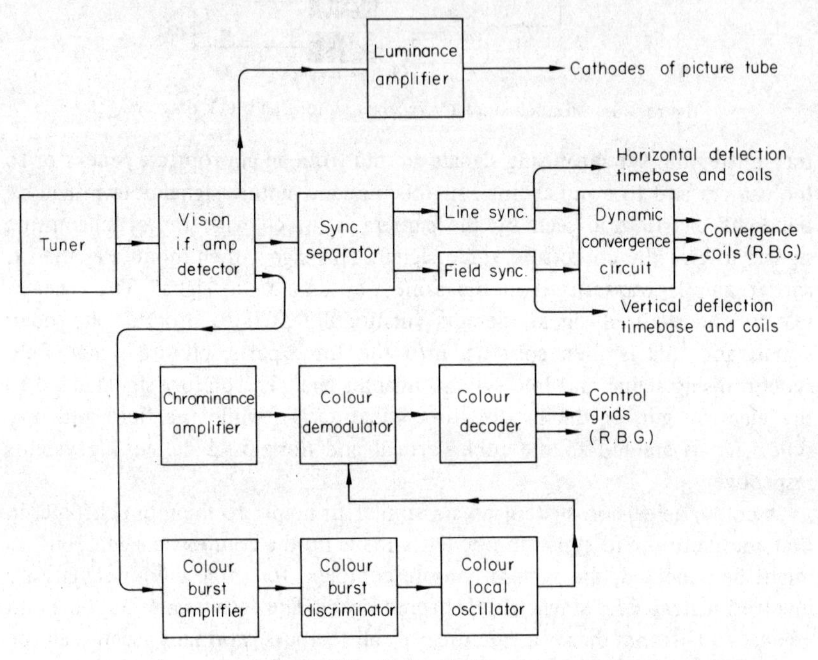

figure 3.9 Colour TV receiver

picture tube. The colour sync. signal acts as an advance warning that a colour signal is coming through, and is contained in a stream of colour *'bursts'*. This signal is amplified and then fed to a local oscillator which together with the colour demodulator helps separate the colour signal into its three component parts. The frequency of the colour local oscillator is synchronised with the colour sub-carrier frequency of the original transmitted signal by the colour discriminator situated between the colour burst amplifier and the colour local oscillator. The picture synchronising signal contains line and field sync. signals as in the monochrome receiver. In addition to being fed to the appropriate parts of the tube deflection systems, they are also fed to a *convergence circuit* which ensures that the three beams are maintained at the correct spacing relative to each other as the beams move over the screen.

3.3. Computer Systems

Figure 3.10 shows a simplified block diagram of a computer. It consists of five subunits: an input section, output section, central processing unit, arithmetic unit and store.

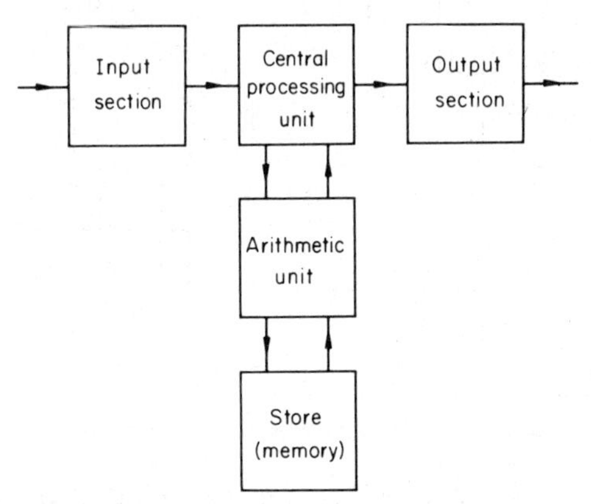

figure 3.10 Simple computer system

As computing systems use coded pulse signals for all their internal operations, information and instructions on how the data is to be processed must first be translated from human language into a language understood by the machine. This is done by the use of an intermediate language which differs from that used in everyday life but nevertheless uses a mixture of English words and symbols (examples of these semi-machine languages include FORTRAN, COBOL and so on, each language being designed to suit a specific purpose). The instructions or *program* given to the machine is therefore first written in an appropriate language, and is then translated into the machine language by the input subunit shown. When the computer has completed the task that it has been instructed to do, the result of that computation is translated from machine language back into everyday language (and decimal figures) by the output subunit. The central processing unit controls the passage of data from the input to-and-from the store, or *memory* (which holds all information and instructions until required), and also controls the arithmetic unit which carries out the required mathematical operations on the given data. The passage of signals throughout the system is controlled by systems of gates (described in the previous chapter and in chapter 9) that are themselves opened and closed by signals from the central processing unit. A slightly more detailed block diagram in figure 3.11 has similar subunits to the previous system and, in addition, has several subunits

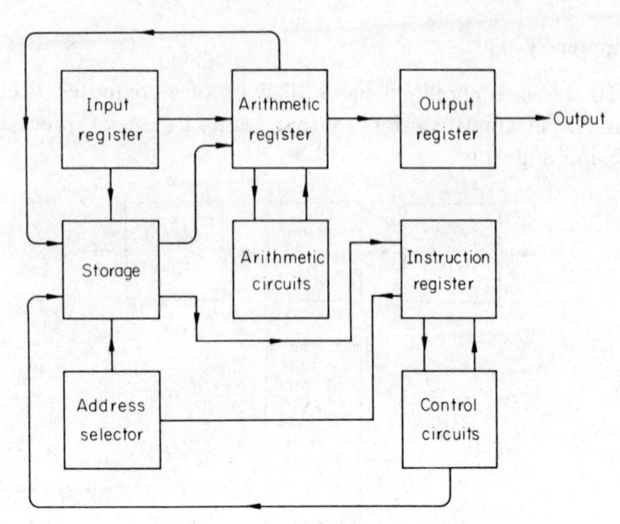

figure 3.11 Computer system with registers

called *registers.* A register is a temporary storage unit that holds information extracted from the main store, or from the arithmetic unit, until the next process is ready to start.

In the system shown, the control unit allows each instruction in turn to be taken from the main store (where they were stored when the program was fed into the computer) to the instruction register. The instruction is then read, and the control unit opens the appropriate gates for the next piece of data to be transferred from the *address* within the memory at which it is being held (the appropriate address is chosen via the address selector subunit) first to the arithmetic register, and then to the arithmetic subunit. The arithmetic register is also used to hold information temporarily at various intermediate stages of the arithmetic process. The result of any process then either goes back into store until the program calls for it to be taken out again, or (on completion of the program) is passed from the arithmetic subunit via the register to the output section ready for read out.

4 Active and passive components

Chapter 2 was concerned with the various types of electronic subunits that make up a system. These subunits are themselves made up of a number of *electronic components* connected together to form electrical current paths called a *circuit.* A diagram showing how the components are connected is called a *wiring diagram* if it shows the exact location of the components and the wires connecting them, and a *schematic diagram* if it shows how the components are connected but does not show the physical location of the components within a subunit. To enable a drawing of this type to be made, symbols are used for components; and, in the interests of general comprehension, these symbols are laid down by the British Standards Institution (BS 3939: Electronic Circuit Symbols).

4.1. Types of Component

Electronic components may be divided into two types: active components and passive components. Active components include electronic *valves, transistors, diodes* and other devices, most of which are capable of amplifying electronic signals. Passive components include *resistors, inductors* and *capacitors* and are not capable of amplifying signals without the aid of an appropriate active device. Before studying the various components it is necessary first to consider basic atomic theory.

4.2. Basic Atomic Theory

It is believed that all materials are made up of basic 'building blocks' called *atoms*; a combination of atoms, for example two hydrogen atoms plus one oxygen atom (H_2O), being called a *molecule.* Atoms themselves are made up of further extremely small parts called *electrons, protons* and *neutrons.* Electrons and protons are electrically charged, and the charges are called 'negative' and 'positive' respectively. The *nucleus* of the atom is composed of protons and neutrons, which are heavy uncharged particles, and the electrons move round the nucleus in *orbits* or *shells.* The distance from the nucleus of these orbits is determined by the electron energy. The number of electrons, protons and neutrons in an atom determine the physical, chemical and electrical properties of the material. All atoms are electrically *neutral* and contain equal numbers of negative electrons and positive protons. If an atom gains or loses an electron it becomes, respectively, either negatively or positively charged, and is then called an *ion.* (See figure 4.1.)

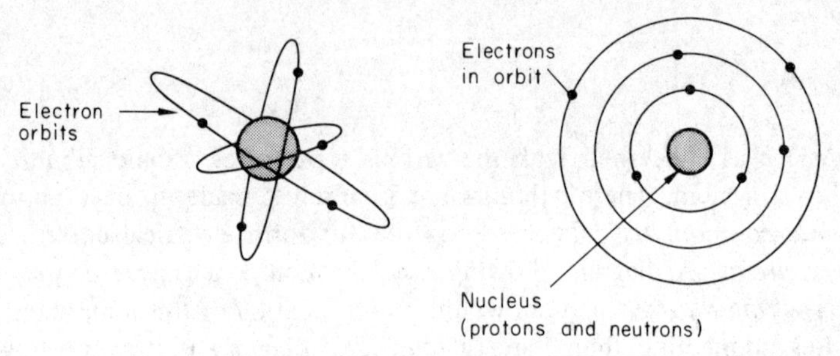

figure 4.1 Basic atomic structure

If electrons are given sufficient energy they may leave their parent atoms and are then free to move through the material, or to leave the material altogether. When electrons move, electric charge is conveyed through the material and this flow of charge is called *electric current.* The unit of electric current is the *ampere,* abbreviated A, where 1A represents a movement of about 6.3×10^{18} electrons every second.

To cause the movement of electrons, and thus set up an electric current flow, an energy source is necessary. This energy may be derived from chemicals (as in batteries), electromagnetically (as in generating machinery), or from heat, light or other radiation. A measure of energy per unit of electric charge is called *voltage,* and this may be expressed variously as electromotive force (e.m.f.) or potential difference (p.d.). The unit of e.m.f. or p.d. is the *volt,* abbreviated V. One volt represents an energy level of *one joule* per *coulomb* of electric charge, where the coulomb represents the charge carried by 6.3×10^{18} electrons. (An idea of the size of the energy unit, the joule, may be obtained from the fact that a one bar electric fire (1 kW) uses 1000 joules every second.)

4.3. Passive Components

The flow of electric current through a material for a particular value of applied voltage depends upon the structure of the material. In some structures a very high voltage is required to set up even a small current; these materials are called *insulators* and are used whenever electric current flow is to be restricted. *Conductors* are materials in which it is fairly easy to establish current, only low voltages being necessary. If the voltage applied to a component is divided by the current through that component the ratio is called *resistance* and it is measured in volts/ampere or *ohms,* for which the symbol is Ω. (See appendix 4 for multiple and submultiple units.) A passive component especially constructed to have resistance is called a *resistor,* and types include *carbon composition, carbon film, metal oxide* and *wirewound* (the name describing the type of material used). Resistors may be fixed or variable, and the construction details of certain types accompanied by their symbols are shown in figure 4.2. Large fixed resistors have their resistance value written on the outside in figures, but the value is usually indicated by coloured rings (see appendix 1 for the colour code used) if they are of the smaller fixed resistance type. An important characteristic to be carefully watched is the *power rating* of a resistor, as components operated at greater power levels than their rating will burn out.

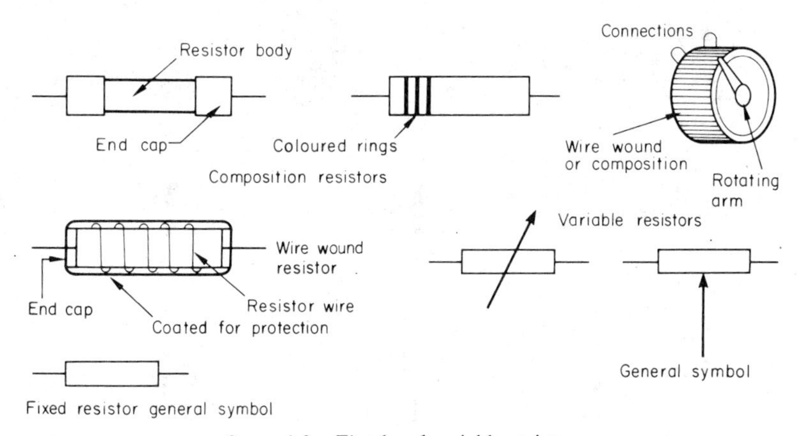

figure 4.2 Fixed and variable resistors

The second passive component to be considered is the *capacitor.* A capacitor is a device that comprises one or more sets of conductive plates separated by a good insulator. Capacitors are specially made to exhibit *capacitance,* which is the ability to store electric charge. A capacitor in a d.c. circuit will charge up, and when fully charged current flow stops. Thus, in normal use a good capacitor has a very high resistance to d.c. and may be used to block d.c. signals. The opposition to alternating current, however, is not as high since the conductive plates alternately charge and discharge. Opposition to a.c. is called *capacitive reactance,* symbol X_c, and it is measured in ohms. Capacitive reactance is given by $X_c = 1/2\pi fC$ where f is the frequency (hertz) and C the capacitance, measured in *farads.* (See appendix 4 for multiple and submultiple units.) One farad (symbol F) means that one coulomb of charge is stored for every volt applied. It follows from the formula just given that a capacitor offers an opposition which *falls* as the signal frequency *rises*; and, consequently, a capacitor may be used to block d.c. but still allow a.c. signals to pass. There are various types of capacitor, including *paper, ceramic, mica* and *electrolytic* (the names describing the insulator used between the plates). Variable capacitors also are available, in most of which moving plates are used to alter the capacitance. For typical construction of fixed and variable capacitors together with their appropriate symbols see figure 4.3. Care must be taken when using capacitors to ensure that their *rated working voltage* (shown on the case) is not exceeded and that the poles of electrolytic capacitors are correctly connected to the d.c. supply.

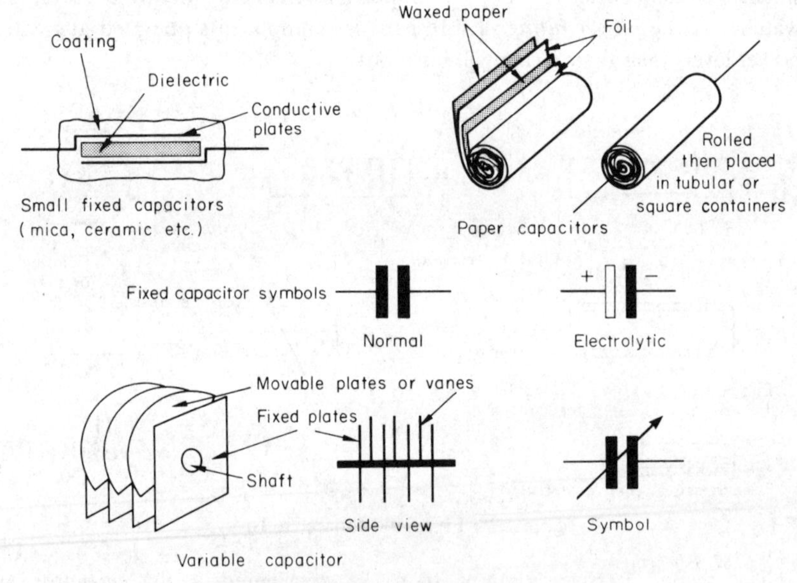

figure 4.3 Fixed and variable capacitors

An *inductor* is a passive component specially designed to have *inductance,* which is a measure of opposition to *changing* current. An inductor connected in a d.c. circuit will slow down the *rate of rise* of current when the switch is closed. Once steady conditions are established the current settles at a value determined by the applied voltage and the inductor resistance. The property of inductance is due to *electromagnetic induction,* which is the setting up of a voltage across a conductor whenever a magnetic field around the conductor is changing. Since any electric current sets up a magnetic field, a *changing* current produces a changing field and thus an induced voltage. This induced voltage tries to stop the change causing it, so that an inductor opposes changing current. As increasing the magnetic field set up by a current increases the induced voltage and therefore the inductance, the inductance can be increased by wrapping the wire in the form of a coil. The field (and inductance) may be further increased by wrapping the coil round a magnetic material such as iron. Thus, high valued inductors are heavy components made up of many turned coils and iron cores. At high frequencies much energy is lost in an iron core and so iron-dust or air cores are used. The inductance of a coil may be varied by moving the core to weaken or strengthen the field. Typical construction and symbols are shown in figure 4.4. The inductance of any coil is measured in *henrys,* symbol H. (See appendix 4 for multiple and submultiple units.) A current changing at the rate of *one ampere per second* in an inductor of *one henry* will set up an opposing induced voltage of *one volt.*

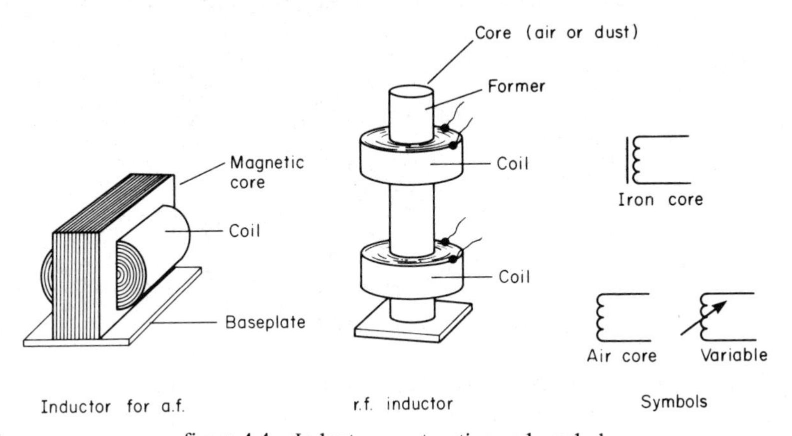

figure 4.4 Inductor construction and symbols

In a d.c. circuit an inductor slows down the change of current on switching *on* and again on switching *off.* In the steady state the current is not changing and is affected only by the resistance of the coil wire. However, a.c. is changing all the time, and thus an inductor offers a continual opposition to a.c. This opposition is called *inductive reactance,* its symbol is X_L, and it is measured in ohms. X_L is related to frequency f (hertz) and inductance L (henrys) by the equation $X_L = 2\pi fL$ and therefore an inductor offers an opposition that *rises* as the signal frequency *rises.* Thus an inductor is used either to oppose the passage of a.c. signals, or else to develop a high voltage between its terminations when an alternating current is passed through it.

An important property of reactive components such as inductors and capacitors is the *phase changing* effect that they have on alternating voltages and currents. In a pure resistance the a.c. flowing when an alternating voltage is applied rises and falls at the same time as the voltage and is said to be *in phase* with the voltage. (See figure 4.5a.) A capacitor, however, because of the time it takes to charge and develop a voltage across it, causes the voltage to *lag* behind the current by 90° or one quarter of a cycle as shown in figure 4.5b. An inductor, on the other hand, because of its current delaying property, causes the current to lag behind the voltage by 90° as shown in figure 4.5c. Thus, we see that inductors and capacitors have 'opposite' characteristics, one offering current lag and a reactance which *increases* with frequency, the other offering voltage lag and a reactance which *decreases* with frequency. This 'opposite' nature is used to good effect in *tuned circuits,* the heart of all radio and television systems. The detailed principles of tuned circuits will not be given here, as this text is devoted to electronic systems rather than electrical principles, but the main points are summarised below.

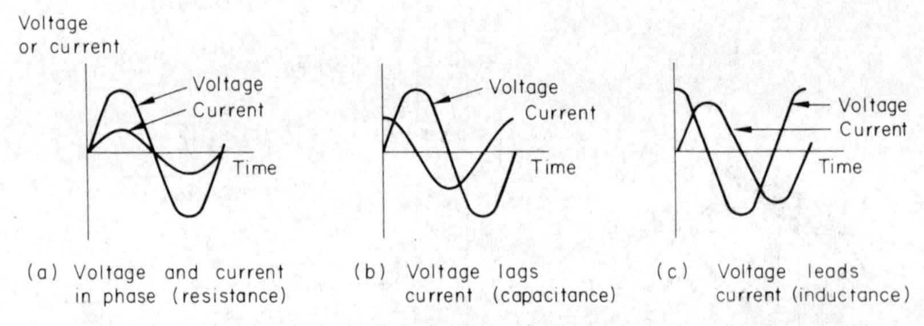

figure 4.5 Phase relationships

If a variable frequency supply is applied to a *series* circuit made up of an inductor, capacitor and resistor in series (see figure 4.6a) the circuit is *capacitive* at *low* frequencies, for at these frequencies the capacitor has more effect than the inductor, and is *inductive* at high frequencies for here the inductor has more effect. At a frequency in between, called the *resonant frequency,* the capacitive and inductive effects cancel and the circuit is purely resistive, thereby allowing a large current to flow (as the total *reactance* is zero).

If a variable frequency supply is applied to a *parallel* circuit comprising an inductor and capacitor in parallel the opposite effect occurs, the circuit being *inductive* at *low* frequencies and *capacitive* at *high* frequencies. Again, at a certain frequency (the resonant frequency) the effects cancel and the circuit

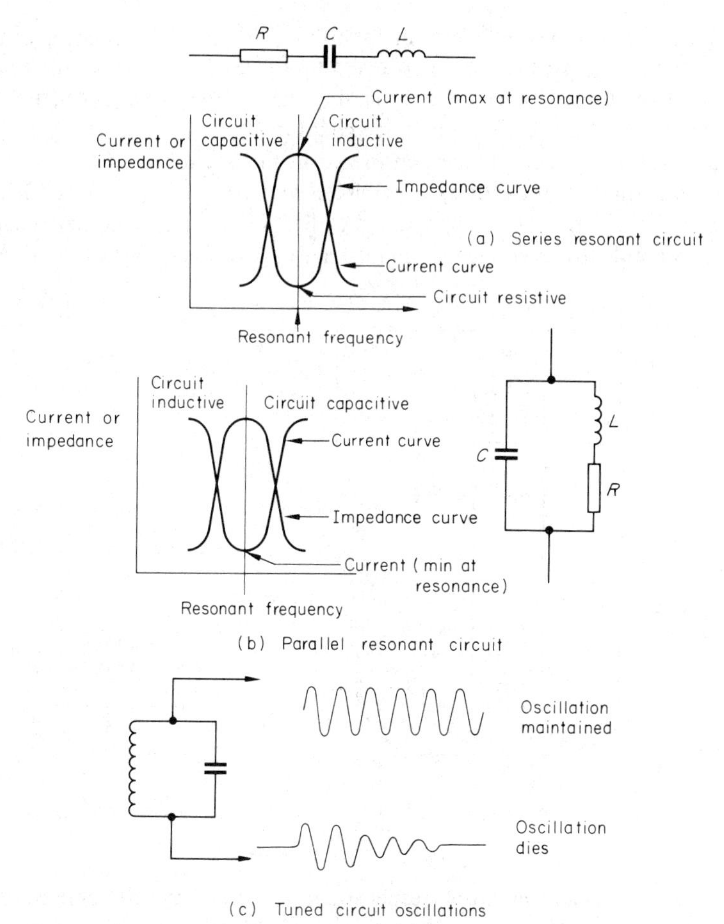

figure 4.6 Resonant circuits

is resistive, but this time the resistance is very high and maximum voltage is developed across the circuit. (See figure 4.6b.) A parallel tuned circuit is a *natural oscillator.* If a voltage is applied across the circuit the capacitor charges to the maximum value. If the voltage is now removed the capacitor discharges and current flows through the coil. The magnetic field builds up to a maximum. The field then begins to collapse (because there is nothing to sustain it) and the capacitor recharges. Thus electrical energy is passed first from capacitor to coil and then back again. If the circuit had zero resistance this oscillating current (shown in figure 4.6c) would continue indefinitely. However, as all circuits have resistance, energy is lost in forcing current through the wires and the oscillations die away. Oscillators using tuned circuits work on the principle of maintaining these natural oscillations, either by feeding energy to the circuit to make up for the loss, or by counteracting the effect of the resistance by a device having 'negative' resistance (such as a tetrode or tunnel diode—for which, see chapter 7).

The final component to be considered in this section is the *transformer.* The transformer is not strictly a passive device because signal amplification can take place within it. As was stated earlier, when a magnetic field changes around a conductor a voltage is induced across it. If (as shown in figure 4.7a)

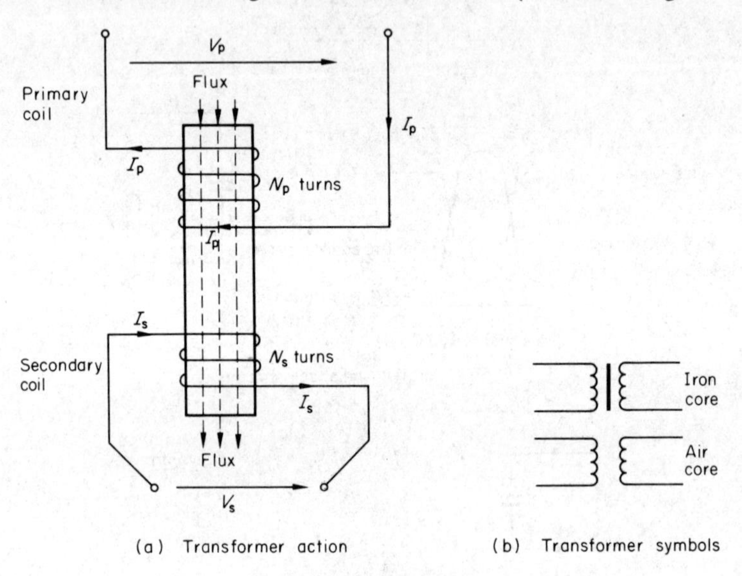

figure 4.7 Transformers

two coils of N_p and N_s turns respectively are linked by the same magnetic field set up by a voltage V_p applied to the coil having N_p turns, a voltage V_s is induced across the coil having N_s turns and the induced and applied voltages are related by the equation $V_s/V_p = N_s/N_p$; which can also be expressed as $V_s = V_p(N_s/N_p)$. Thus if a turns ratio (N_s/N_p) of 10/1 is used, 10 V may be induced in the secondary for every 1 V applied to the primary. (The coil across which the applied voltage is connected is called the primary coil; the other, the secondary coil.) The secondary current I_s is related to the primary current I_p by the equation $I_s/I_p = N_p/N_s$; which can be re-written as $I_s = I_p(N_p/N_s)$. For a 10/1 secondary/primary turns ratio, the secondary current would be $\frac{1}{10}$ of the primary current. *Electrical power* is the product of voltage and current. It can be seen that V_s multiplied by I_s is $V_p(N_s/N_p)$ multiplied by $I_p(N_p/N_s)$, which cancels down to V_pI_p; from which it is clear that the output power (V_sI_s) is equal to the input power (V_pI_p).

Transformers are widely used in electronic systems for changing voltage and current levels and for *matching* one load to another (see appendix 2). High power transformers use many-turned coils and heavy iron cores (to increase the linking magnetic field). At high frequencies core losses are high, so dust cores or air cores are used as with inductors. Transformer symbols are shown in figure 4.7b.

4.4. Active Devices

Active devices may be divided into two types: gas filled or vacuum devices and solid state devices. In the first type, the electronic *valve*, the flow of electric charge carriers is confined to a vacuum or to an inert gas at low pressure. In the second type current flow takes place within a specially prepared solid called a *semiconductor*. All active devices have two or more points to which connections are made and these are called *electrodes*. A two electrode device is called a *diode*, a three electrode device a *triode*. In addition, there is a *tetrode*, *pentode*, *hexode* and *heptode* with four, five, six and seven electrodes respectively.

4.5. Diodes

A diode has two electrodes called the *anode* and *cathode*. The resistance offered by a diode to the flow of electric current is very low in one direction and very high in the other direction. The diode is thus a *unidirectional* device. To obtain maximum current in the low resistance or *forward* direction, the anode must be at a positive potential with respect to the cathode. A graph of voltage against current has the shape shown in figure 4.8a for a vacuum diode valve and 4.8b for a semiconductor diode. Notice that the diode valve allows no current flow in the reverse direction but the semiconductor diode allows a very small current flow up to the point of breakdown. Breakdown in solid state diodes is discussed in more detail later.

Vacuum diodes work on the principle of electrons being released from the cathode and then being attracted to the anode whenever that electrode is positive with respect to the cathode. If the anode is at a negative potential with respect to the cathode the electrons do not move to the anode and current flow is zero. Electrons are released from the cathode by heating it, using a high resistance *filament* through which *heater current* is passed. The use of a heater in this manner is the origin of the name *thermionic* valve.

In directly-heated valves the filament is itself the cathode and must be of a material that can stand high temperature and that is capable of emitting a copious supply of electrons. A typical material is tungsten, which emits

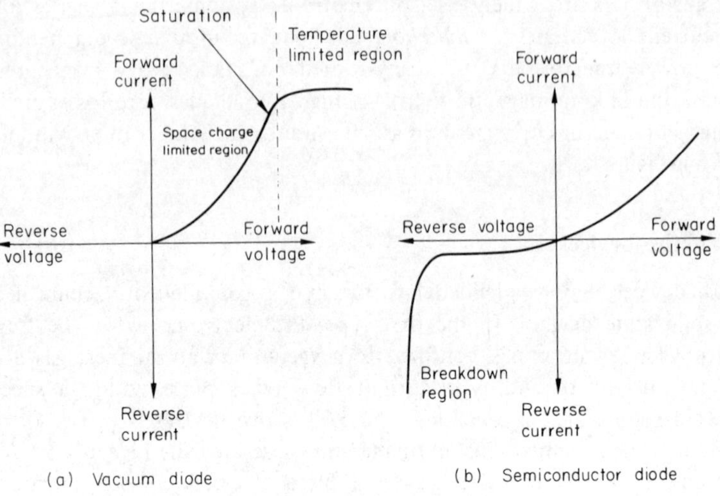

figure 4.8 Diode characteristic

electrons at a temperature of about 2300° C. Coating the tungsten cathode with a material such as thorium, for example, allows increased emission at a lower temperature of about 1700° C. In indirectly-heated valves the cathode takes the form of a metal sleeve fitted over the heater filament (as shown in figure 4.9a) and coated with materials that emit electrons at much lower temperatures (around 700° C), typical examples being the oxides of barium, calcium and strontium. In both types the anode is in the form of a cylinder and surrounds the cathode completely as shown in the figure. Circuit symbols for use on a schematic circuit diagram are shown in figure 4.9b.

The diode characteristic shown in figure 4.8a is in two parts. The first shows that after an initial *non-linear* increase, the current rises approximately *linearly* as the anode-cathode voltage is increased. In the second part the current remains approximately constant as the voltage is increased. With no applied voltage the electrons emitted by the cathode tend to gather in a dense cloud called the *space charge.* The space charge remains fairly constant, because as some electrons are emitted others lose their energy and fall back into the cathode so that a state of equilibrium exists. As the anode voltage is increased electrons are attracted across to the anode, the number (and therefore the current) increasing in approximately direct proportion to the voltage. The region of the curve where this happens is called the *space-charge limited region.* Eventually, all electrons travel across as fast as they are emitted and the current *saturates* at a value determined by the cathode temperature. The graph now enters the *temperature-limited* region. The only

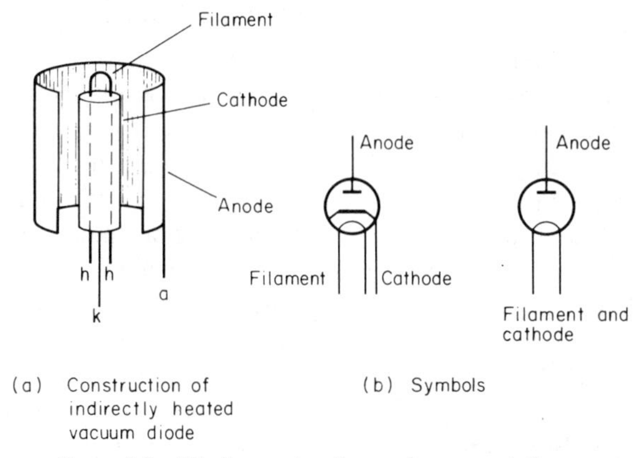

figure 4.9 Diode construction and representation

way to increase current in this region is by increasing the heater current and thus the cathode temperature. There is, of course, a limit to the working temperature, and this is usually specified by the manufacturer. The circuit used to obtain the anode-current anode-voltage graph shown earlier is given in figure 4.10. A variable voltage is applied between anode and cathode and the valve current read on the ammeter. Notice that the heater supply is separate from the anode-cathode supply.

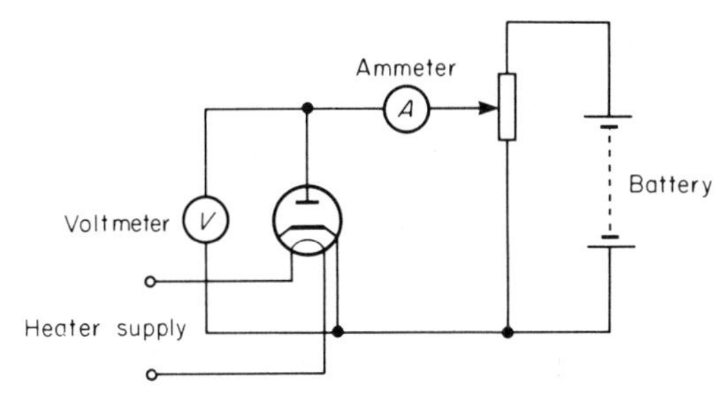

figure 4.10 Circuit to obtain *IV* curve

Semiconductor diodes work on an entirely different principle to that of thermionic valves. Pure semiconductor materials, the commonest of which are *germanium* and *silicon,* are materials having an electrical resistivity lying between that of good conductors and insulators. (Note: *Resistivity* of a material is the opposition to current flow presented by a standard unit of that material, the unit normally specified being the opposite faces of a cube of side length one metre.) In the pure or *intrinsic* form semiconductors have a fairly high resistance and allow only a very small current flow. With the addition of certain impurities, however, the conducting ability can be altered considerably. The basic principles are as follows.

Pure semiconductors are what is known as *tetravalent.* This means that each atom of the material shares four of its electrons with its neighbouring atoms, as illustrated diagrammatically in figure 4.11a. A pentavalent material is one whose atoms share *five* electrons and trivalent material is one whose atoms share *three* electrons. If a very small amount of a pentavalent material, for example arsenic, is introduced into the semiconductor crystal (about 1 part in a million) the pentavalent atoms share *four* of the available *five* with the neighbouring semiconductor atoms and thus one electron per pentavalent atom is unshared. Shared electrons are very tightly bound within the crystal and are not easily moved. This fact explains the high value of resistivity of an intrinsic crystal. However, with the pentavalent impurity present one electron per pentavalent atom is not so tightly bound and is thus available to take part in current flow. A semiconductor crystal *doped* with impurity in this manner has a much reduced resistivity. The doped crystal is called *extrinsic* semiconductor and is known as N-type, the N standing for negative which is the nature of the charge carried by an electron. (See figure 4.11b.)

If a similarly small amount of a trivalent impurity such as indium or gallium is introduced into a semiconductor crystal only three electrons are available for sharing and the normally tetravalent bond now contains a gap shown symbolically in figure 4.11c. This gap, known as a 'hole', acts as a trap

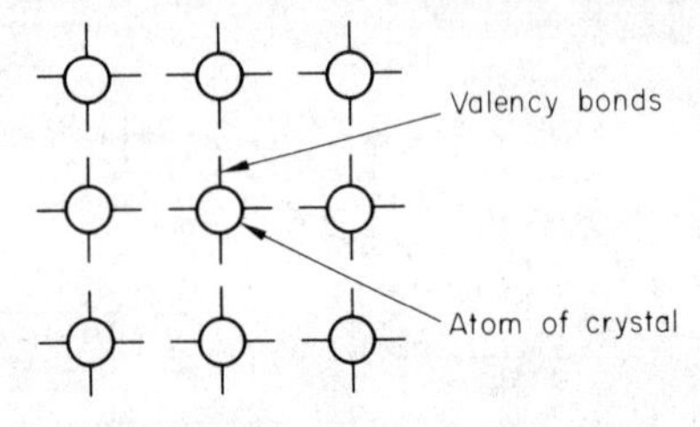

(a) Intrinsic (pure) semiconductor

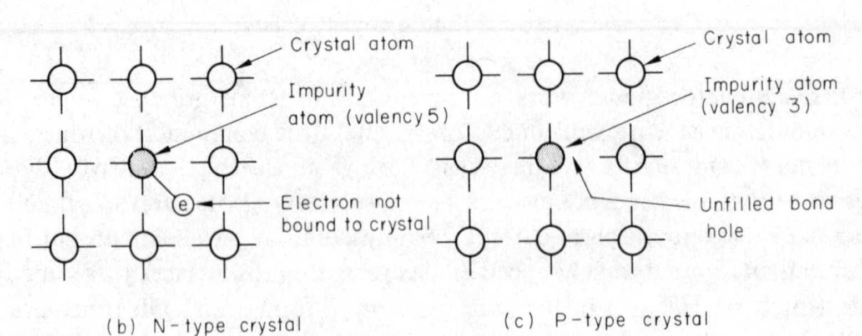

(b) N-type crystal (c) P-type crystal

figure 4.11 Crystal doping

for electrons from neighbouring atoms. An electron from a tetravalent bond associated with the atom containing the hole can fill the hole if the electron receives sufficient energy to leave its bond. As the hole is adjacent, the energy required is not as much as that required for an electron to leave a tetravalent bond altogether. When a valency electron moves in this manner it leaves a new hole behind it and the hole effectively moves through the material. The process can then continue, the hole moving from one atom to an adjacent one. Thus a current flow is established more easily within the crystal and its resistivity falls. A crystal doped with trivalent material is called P-type since the hole, being attractive to an electron, may be regarded as a *positive* charge. Hole conduction is illustrated in figure 4.12.

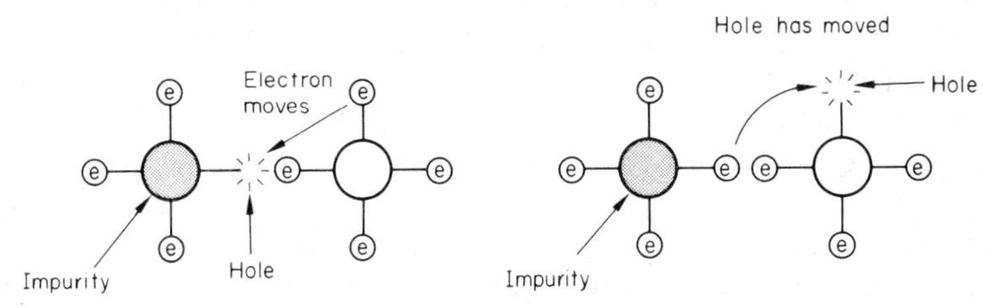

figure 4.12 Hole conduction

P-type and N-type extrinsic crystals are electrically neutral even though they have extra charge carriers available for conduction. The reason for this is that each atom of impurity is itself neutral, the intrinsic material is neutral and so the doped crystal is also neutral. The reason that the charge carriers are available for conduction lies in the tetravalent nature of the semiconductor and the fact that the bond is either over-filled or under-filled. A junction of P-type and N-type material, however, causes the neutral condition to disappear and the junction can be used as a diode. This may be explained in the following manner.

Consider a piece of doped semiconductor that is P-type at one end and N-type at the other as shown in figure 4.13a. At room temperatures the presence of heat energy causes the holes in the P-type and the electrons in the N-type material to move about randomly as determined by their respective energies. Some electrons cross the junction from the N-type to the P-type and are 'captured' by the holes. The P-type which was neutral becomes negative and the N-type which was neutral, having lost electrons, becomes positive. Thus a potential difference, the *junction p.d.* is set up across the junction. This p.d. eventually stops the random drift of electrons since the now negative P-type repels them. If now an external battery is connected across the junction, positive to N-type, negative to P-type (as shown in figure 4.13b)

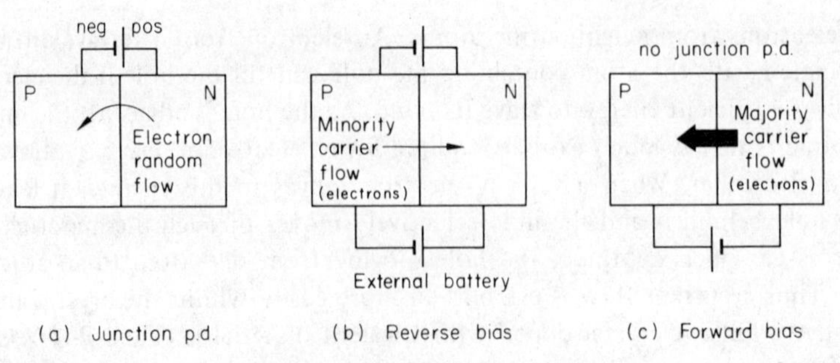

figure 4.13 The PN junction

the junction p.d. is reinforced and the N-type electrons do not cross the junction. This connection is called *reverse bias.* Unlike the valve diode in this condition (that is, with anode negative relative to the cathode) current does not stop altogether. Although each type of doped semiconductor contains a majority of one kind of charge carrier (that is, N-type with electrons and P-type with holes) *minority carriers* exist in both types: N-type contains some holes and P-type contains some electrons. This is the result of valency electrons absorbing heat energy from the surroundings and leaving their bonds, thus leaving holes behind them. The process is known as *thermal generation* of hole-electron pairs. Thermal generation takes place at all temperatures above absolute zero. When the junction is reverse-biased, minority electrons from the P-type cross the junction and travel to the positive side of the supply voltage. Thus, a small current flows. This current is sometimes called saturation current because it is temperature controlled in the same way as the valve diode anode current is temperature controlled when saturated. Notice, however, that saturation in the semiconductor is in the *reverse* direction; in a valve diode there is no reverse current. If the reverse bias external voltage is increased beyond a certain limit, determined by the physical characteristics of the junction, *breakdown* occurs and the reverse current rises to a value comparable with that in the forward direction. Breakdown may take one of two forms: *zener breakdown* or *avalanche breakdown.* Zener breakdown occurs because the intense electric field causes generation of hole-electron pairs as a result of valency electrons absorbing energy. Avalanche breakdown is caused by minority carriers travelling at high speed through the reverse field and dislodging electrons which are then themselves accelerated and thereby dislodge further electrons. The type of breakdown which occurs depends upon the sharpness of the junction between the P-type and N-type materials. Zener breakdown occurs in junctions where the change is abrupt, that is, the junction region is narrow; avalanche breakdowns occur in more gradual or broader junctions. This part of the current-voltage characteristic is used in voltage regulator diodes (described later).

If an external battery is applied as shown in figure 4.13c (that is, positive to P-type, negative to N-type) the internal junction p.d. set up by random carrier movement is neutralised and *majority carrier* current flows. This means that N-type electrons flow through the junction to the positive side of the supply and P-type holes move through the junction to the negative side of the supply. It should be noted that in semiconductor current flow electron current flow is *added* to the hole current flow to give total current. Electron current flow is caused by high energy electrons moving freely through the crystal, hole current flow is caused by low energy electrons moving from bond to bond. With the supply voltage connected in this way, called the *forward-bias* condition, a fairly large forward current flows as shown in figure 4.8b.

The circuit used for obtaining the current-voltage curve of a semiconductor diode is similar to that for the vacuum diode shown in figure 4.10 with the voltages and polarities suitably adjusted and re-arranged. Semiconductor diodes normally operate at much lower voltages than their valve equivalents. Both kinds of diode are used for similar purposes—rectification, detection, etc.—and the operating conditions and requirements are discussed in subsequent chapters. However, semiconductor diodes have distinct advantages over valves: they are much smaller; they do not require a filament supply; and because of their construction they are more robust.

As stated, the main uses of the diodes discussed so far are in rectification and detection. In addition to these general purpose diodes, certain other diode devices, such as voltage and current regulator diodes, varactor diodes and tunnel diodes, are used for special purposes.

A *voltage regulator* or *stabiliser diode* is used to maintain an approximately constant voltage across a load over a wide range of currents taken by that load. A suitable component to do this is a P–N junction diode operated in the breakdown region of the current-voltage characteristic. (See figure 4.8b.) Such a diode is, of course, specially designed with adequate junction width to permit continuous operation in the breakdown region. Examination of this region shows that the voltage is approximately constant over a range of reverse currents. The reverse current is small compared to normal forward current, but suitable circuits can be used to increase the available load current (chapter 5). A diode operated in the breakdown region is called a *Zener diode.* An *avalanche diode* is a similar device but uses avalanche breakdown, as described earlier. The valve equivalent of the Zener diode is the *cold-cathode gas-filled diode,* which has a current-voltage curve as shown in figure 4.14. This diode valve contains a small proportion of an inert

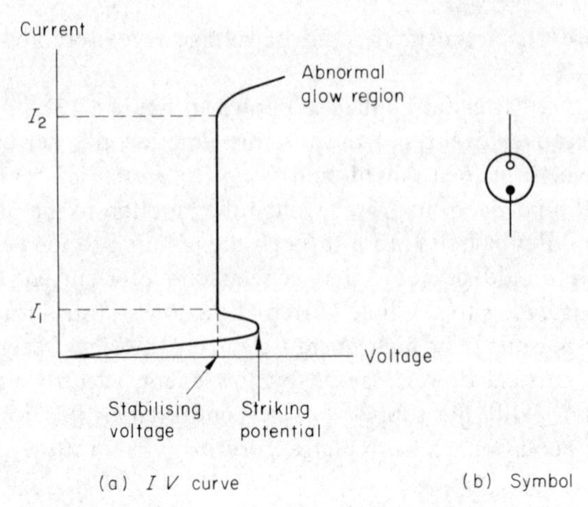

figure 4.14 The cold cathode-valve

gas at low pressure. No heater is necessary. As the anode-cathode p.d. is increased a small current flows due to the free electrons contained within the envelope. At a certain voltage, the *striking potential,* these electrons are travelling towards the positive anode at sufficient speed to dislodge further electrons from the gas atoms, and this effect is *cumulative* (as in the case of semiconductor avalanche breakdown, described earlier). The gas atoms, having lost electrons, become *positive ions* and move towards the cathode. Impact with the cathode causes a release of further electrons, which add to the current. One point to note about the diode stabiliser curve of figure 4.14a is that the stabilising voltage is slightly less than the striking voltage, because the anode-cathode resistance drops once the diode has struck. The valve should normally be operated between the two current limits laid down by the manufacturer. Below the lower limit, I_1 on the curve, the valve extinguishes; above the upper limit, I_2 on the curve, the valve enters the *abnormal glow* region and damage may result. During normal operation the colour of the valve glow is determined by the gas content. The symbol for a cold-cathode stabiliser is shown in figure 4.14b.

In the *current regulator* or *constant current diode,* the diode current remains approximately constant for varying voltage. The solid-state version uses a *field effect transistor* (discussed in the next article) with *gate* and *source* connected together. The valve version is called a *barretter* but its use is now too limited to warrant inclusion here.

A *varactor diode* is a solid-state device having a *capacitance* which depends upon the applied voltage. Capacitance is the ability to store charge; it is also the ratio between current and *rate of change* of voltage in a circuit. The larger the value of capacitance the longer it takes for a particular voltage change. A passive device specially made to have capacitance is called a *capacitor* (as discussed in article 4.3); but the capacitance of such a capacitor, unlike that of the varactor diode, is not controlled by applied voltage. The varactor diode is a P–N junction diode suitably constructed for continuous operation in the reverse bias condition (but *not* near breakdown as in the case of the Zener diode). Under this condition majority carriers do not flow across the junction as was described earlier. In fact, they tend to move back from the junction, under the influence of the reverse field, and a region situated either side of the junction becomes *depleted* of charge carriers, with the result that its resistivity rises accordingly. The width of this *depletion region* is determined by the reverse field and thus by the reverse voltage. This provides a piece of material that has two end regions separated by what is almost an insulating region, an arrangement very similar to the construction of a capacitor. However (as already stated), this device differs from the normal passive component in that the applied voltage, which varies the thickness of the 'insulating' layer, varies the capacitance.

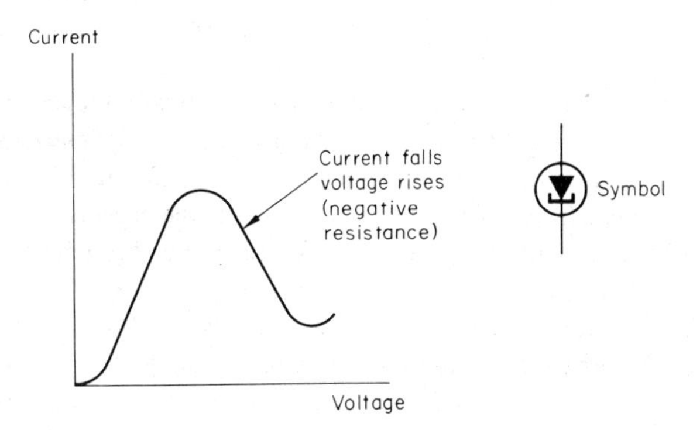

figure 4.15 Tunnel diode *IV* curve

Another semiconductor diode of importance is the *tunnel diode,* which has a current-voltage curve as shown in figure 4.15. As can be seen, as the applied voltage rises the diode current rises then falls and the effect is that of a 'negative' resistance. This is the opposite effect to that displayed by a resistor, for in a resistor current always rises with increasing voltage. Such a component may be used to oppose the effects of positive resistance; as, for example, in a tuned circuit, so that natural oscillations that would otherwise die out can be maintained. The tunnel diode oscillator is described in chapter 7.

4.6. Triode Valves and Transistors

The triode valve has *three* electrodes; a three electrode semiconductor device is called a *transistor.* In the valve the third electrode is a *grid* and it is situated between cathode and anode. The grid consists of a mesh of fine wire as shown in figure 4.16a; the symbol for a triode valve is shown in figure 4.16b.

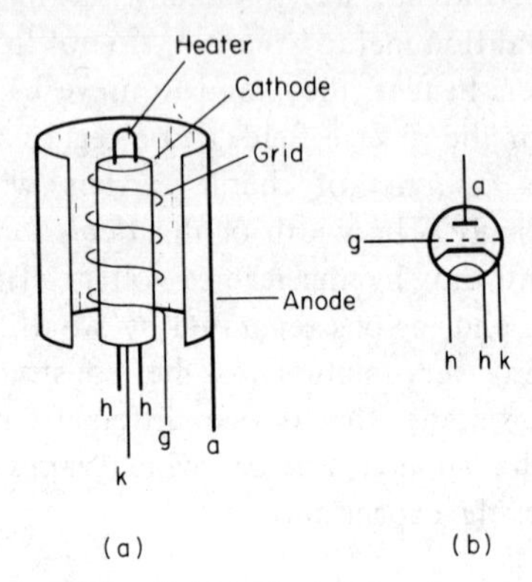

figure 4.16 The triode valve

The current flow in a valve diode consists of a stream of electrons moving from cathode to anode, the number at any one time being determined by the anode–cathode p.d. if the valve is being operated in the space-charge limited region, and by the filament (heater) temperature if the valve is being operated in the temperature limited region. It is found that a fine wire mesh placed between cathode and anode has a considerable influence on the anode current and that this influence depends upon the potential of the mesh with respect to the cathode. When the grid is made more negative with respect to the cathode the valve current is reduced, as the electrons are repelled by the grid field and only the high energy electrons can get through to the anode. Eventually, as the grid-cathode voltage is further increased (negatively), even these cannot get through and the valve is said to be *cut-off.* If the grid is at a positive voltage with respect to the cathode it will attract the electrons and rob the anode of some of its supply. In this case the *total* valve current (cathode current) is increased but the anode current is reduced. A triode is normally not operated with the grid too positive for any length of time, because there is a danger of destroying the fine wire mesh. There are *three* variable quantities for the triode compared with two for the diode. These are anode-cathode voltage, anode current and grid-cathode voltage; the symbols usually employed for these quantities being V_A, I_A and V_G respectively (if *direct* voltages and currents are being considered). As there are three quantities, three possible sets of characteristics can be drawn: I_A plotted against V_A with V_G constant; I_A plotted against V_G with V_A constant; and V_A plotted against V_G with I_A constant. The last of these is rarely used, but the I_A/V_A curve (called the *anode characteristic*) and the I_A/V_G curve (called the *mutual characteristic*) are in frequent use for design purposes. Typical curves are shown in figure 4.17a and b and the circuit for obtaining them in figure 4.17c. As these curves show the operation of the valve without a load, they are called *static* curves.

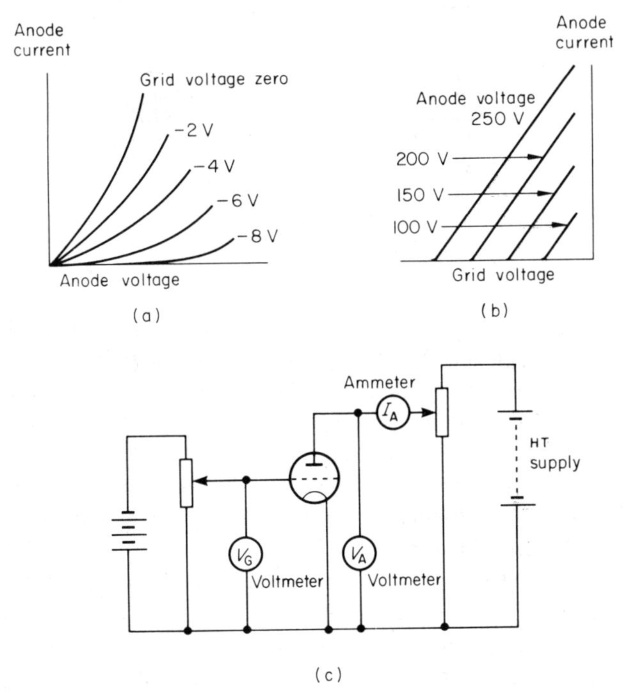

figure 4.17 Triode valve characteristics

Because of the nearness of the triode grid to the cathode its influence (per volt) on the valve current is far greater than that of the anode. If a circuit is so arranged that a change in grid voltage causes a change in anode current which in turn causes a variation in anode voltage, then the anode voltage change is always *larger* than the original grid voltage change that caused it. Consequently, amplification is possible. The circuits used in amplifiers are considered in detail in chapter 6.

The semiconductor component with three electrodes is called a transistor. There are two types: the *bipolar* transistor, which uses both electron and hole conduction; and the *unipolar* or *field-effect* transistor (FET), which uses either electron *or* hole conduction. (It should be remembered that hole conduction also uses electrons as charge-carriers but that these electrons are those normally in the bond between atoms; electron conduction uses 'free' electrons, which do not necessarily move from one atom to the one adjacent.)

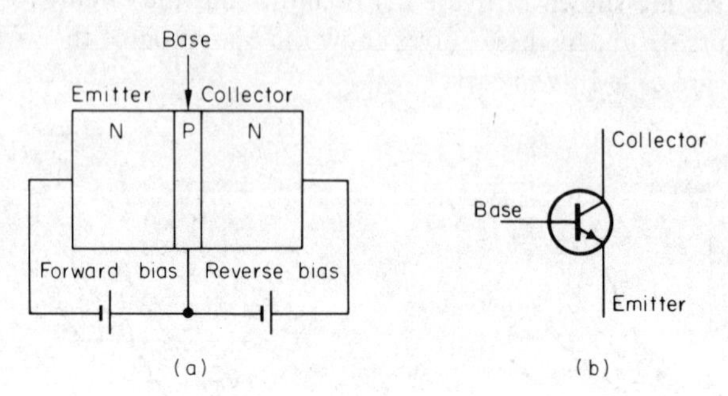

figure 4.18 NPN transistor

A bipolar junction transistor, shown diagrammatically in figure 4.18a, has three layers of doped semiconductor material, the outer layers being of the same type. This permits two choices of layer arrangement which are referred to as P–N–P or N–P–N transistors. The device shown in the figure is an NPN type, the symbol being shown in figure 4.18b. On the PNP symbol the arrow head is reversed. The three layers are called the *emitter, base* and *collector* and may be likened approximately to the cathode, grid and anode of a triode valve. As can be seen there are two PN junctions in the device. For correct working the emitter-base junction is forward-biased and the base-collector junction is reverse-biased as shown in the figure. Due to the forward-bias, majority carriers (electrons for an NPN device) are swept from emitter to base. The base is made very narrow so that the bulk of the majority carriers continue across the base-collector junction to the collector. The collector voltage is of the opposite polarity to the carriers (held positive in the case of the NPN device) and the carriers are thus attracted to it. Some carriers do leave the base terminal and a small base current is set up. Denoting collector, emitter and base currents by I_C, I_E, I_B respectively, the equation

$$I_E = I_C + I_B$$

is true at all times. If I_B is changed by changing the emitter-base forward bias, I_C and I_E change accordingly and it is found that the change in collector current is greater than that of the base current. Thus *current amplification* takes place between base and collector. Using appropriate circuitry, voltage signals can be produced by these currents and the bipolar transistor can then be used in a similar way to that of a triode for voltage amplification.

With the bipolar transistor the important variable quantities are: collector current; collector voltage; and either, voltage between base and emitter, or base current. These are indicated using symbols I_C, V_C, V_{BE} and I_B if direct voltages and currents are being considered. The characteristics of the bipolar transistor that are directly comparable to those of the valve are I_C/V_C with V_{BE} constant, and I_C/V_{BE} with V_C constant. In addition, as the bipolar transistor has a base current, we may plot I_C/V_C with I_B constant and I_C/I_B with V_C constant. Typical graphs are shown in figure 4.19. Notice the difference between the shapes of the collector characteristics and those of the anode characteristics of the triode valve (figure 4.17a). The significance of the shape and of the curves in general is discussed in more detail in chapter 6.

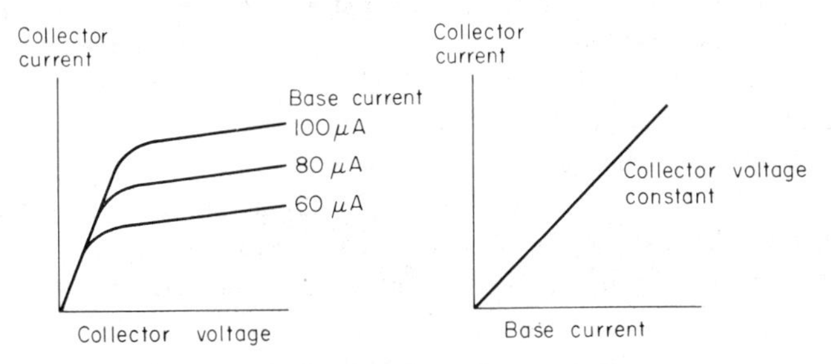

figure 4.19 Transistor characteristics

The field-effect transistor works on rather a different principle to the bipolar transistor. That principle is that if the cross-sectional area of a conductive channel is altered then the channel resistance to current flow is altered, and thus the current obtainable for a given voltage is also altered. If the channel resistance can be altered by some other voltage this is an equivalent arrangement to that of a voltage changing a current level in the valve. The basic layout of a *junction-gate FET* or JUGFET is shown in figure 4.20a. The channel in this version is an N-type channel in a P-type block, but P-type channels in N-type blocks are equally available. Electrons flow through the channel from *source* to *drain,* these electrodes being similar to cathode and anode for a valve, or to emitter and collector for an NPN bipolar transistor. The electrode that controls the channel width, and thus the through-current, is called *the gate.* The gate is similar to the grid of the valve, or to the base of the bipolar transistor. If the P–N junction between gate and channel is *reverse biased,* a depletion region exists each side of that junction and extends into the channel. A depletion region is one in which the density of charge carriers (in figure 4.20, electrons) is reduced. By adjusting the

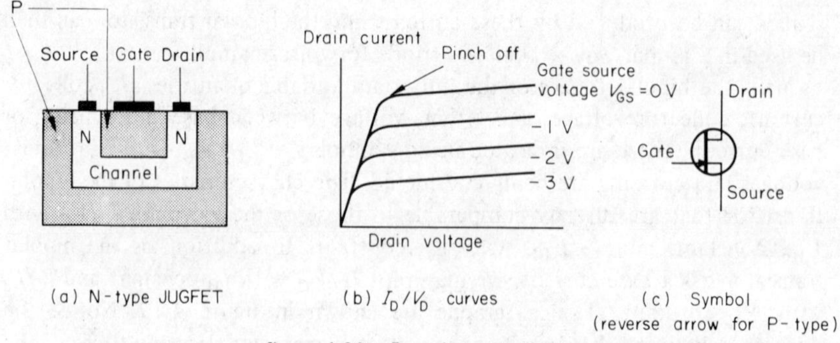

figure 4.20 **Junction gate FET**

gate-channel reverse-bias the channel resistance to current flow may be altered. For a particular gate voltage the reverse bias between gate and channel is not constant all the way along the channel, because of the voltage drop along the channel. The drain end of the channel is at a higher potential with respect to the gate than is the source end of the channel. Thus, the difference between the gate potential and channel potential (that is, the gate-channel reverse-bias) is greater at the drain end of the channel. The channel width is thus more greatly affected at the drain end. For a particular value of gate-source voltage, as the drain voltage is increased the drain current will rise approximately linearly until a point is reached where the tendency for the drain current to increase in response to increasing drain voltage is balanced by the tendency for the drain current to reduce as a result of the narrowing of the channel caused by the gate-channel p.d. At this point, called *pinch-off,* the current *saturates* at an approximately constant value. If the drain-source voltage is further increased beyond this saturation range, avalanche breakdown occurs and the transistor may suffer permanent damage. A family of drain-current, drain-voltage curves, with the gate-source voltage held constant for each curve, is shown in figure 4.20b. Note that the symbols for the FET variable quantities drain current, drain voltage and gate voltage are I_D, V_{DS} or V_D and V_{GS} or V_G if direct quantities (d.c.) are being discussed. The fact that the control per volt over the drain current is greater by the gate-source voltage than by the drain-source voltage makes it possible for the JUGFET to be used for amplification in a similar manner to that for a triode. This is discussed in chapter 6.

A second type of field-effect transistor is the insulated (or isolated) gate field-effect transistor, abbreviated IGFET. In this type the metal gate connection is insulated from the semiconductor block by an oxide layer. The metal–oxide–semiconductor arrangement gives rise to the other abbreviation in common use which is MOSFET. Because of the necessity for extremely thin oxide layers this type of FET is a fairly recent development. There are two types of IGFET, the *depletion-mode* and the *enhancement-mode.* The depletion IGFET works in a similar way to the JUGFET in that a channel exists between source and drain and the channel width is controlled by the gate. The gate control is different, however, because the controlling voltage is not connected directly to the block. For the N-type IGFET shown in figure 4.21a a negative gate voltage *induces* a positive layer directly beneath the

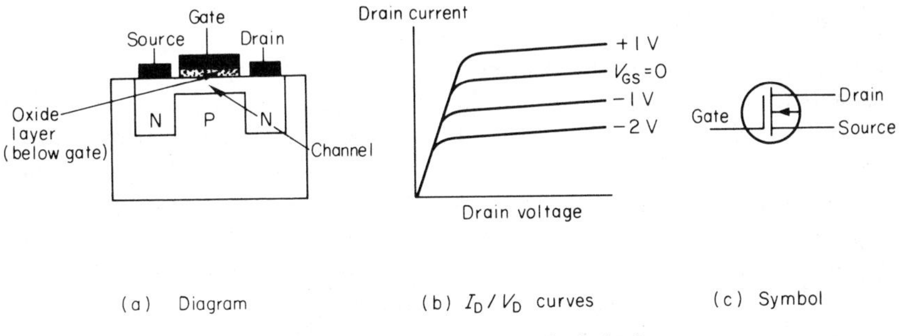

figure 4.21 N-type IGFET (depletion)

gate, by driving electrons from the region. This *depletes* (reduces) the number of charge carriers and thus the current in the channel. Application of a positive gate voltage has the opposite effect and channel current increases. This latter point is a rather different feature from the JUGFET or triode valve in which the control voltage is usually of only one polarity (negative for an N-type JUGFET and triode valve, positive for a P-type JUGFET). A set of I_D/V_D characteristics for an N-type depletion IGFET is shown in figure 4.21b. For a P-type, gate and drain polarities are opposite to those for an N-type, being positive and negative respectively, with respect to the source.

A cross-section of an enhancement IGFET is shown in figure 4.22a, with

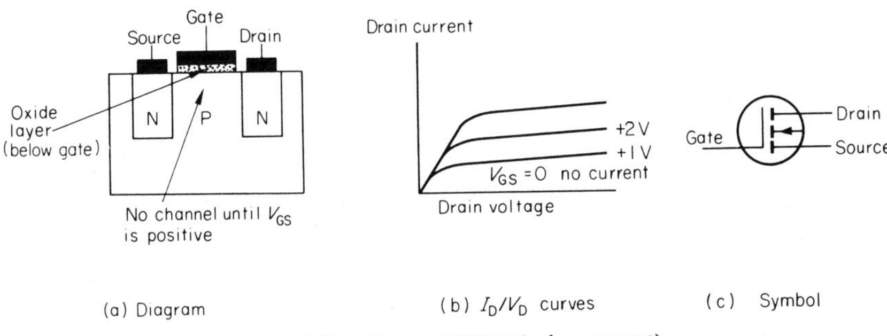

figure 4.22 N-type IGFET (enhancement)

I_D/V_D characteristics in figure 4.22b. In the enhancement-mode device the gate voltage *creates* the channel in the semiconductor block, a positive gate voltage inducing an effective N-type layer between source and drain. With no gate voltage applied current does *not* flow between source and drain. Both types of IGFET may be used as amplifiers because, just as with the JUGFET and triode valve, the control voltage (gate or grid) controls the through current (drain or anode).

4.7. Multi-electrode Valves

One disadvantage of the triode valve is the *interelectrode capacitance* that exists between anode and grid. Capacitance (which was discussed in article 4.3 on passive components) offers an opposition to alternating current flow which falls off with increasing frequency. At higher frequencies, especially in the radio-frequency range, the anode-grid capacitance offers a *low impedance* path to a.c. which is equivalent to a connection between anode and grid. If the output signal is taken from the anode and input signal put in at the grid there is a *feedback* path between output and input. Under certain circumstances (as was pointed out in chapter 2), this can lead to oscillation, which in an amplifier is not desirable. The inclusion of a second grid, the *screen* grid, placed between the first grid and anode, considerably reduces the effect of the anode-grid capacitance. The valve containing the two grids now has four electrodes (anode, cathode, control grid and screen grid) and is called a *tetrode.* A typical anode characteristic of such a valve is shown in figure 4.23. As can be seen, the anode current rises and then falls and then rises again as anode voltage is increased. The unusual fall in I_A with increasing V_A is due to *secondary emission.* At a particular value of V_A the electrons as they strike the anode have sufficient energy to dislodge further electrons. This occurs in a triode but there the dislodged or *secondary* electrons merely move out from the anode and then, under the influence of the anode voltage move back again and there is no effect on anode current. In the tetrode the

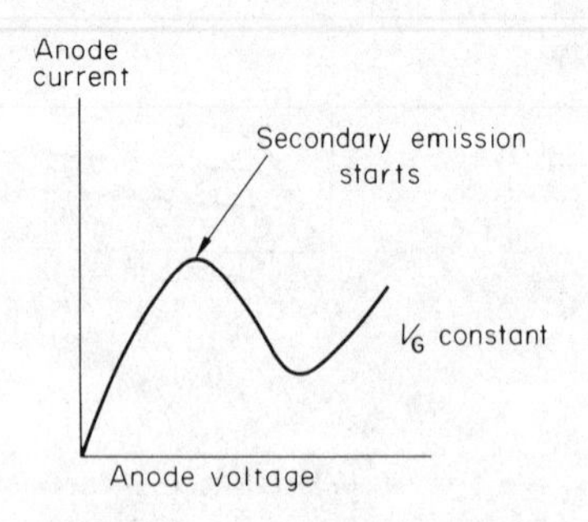

figure 4.23 Tetrode I_A/V_A curve

screen grid, which is normally held positive with respect to the cathode, attracts these secondary electrons and the screen-grid current rises, thereby producing a fall in anode current. As V_A is further increased the anode potential overcomes the effect of the screen grid voltage, which is held constant, and secondary electrons do not reach the screen but return to the anode. The problem is overcome by one of three methods:

(1) by arranging the electrodes at suitable separations from one another and with the screen grid in the 'shadow' of the control grid
(2) by using beam plates
(3) by inserting a third grid, called a *suppressor.*

Method (1) modifies the valve internal electric field pattern and electrons do not reach the screen grid. The valve so produced is called a 'critical distance' tetrode and the I_A/V_A curve 'kink' is removed. Method (2) produces a valve called a *beam tetrode,* in which, plates are inserted (as illustrated in figure 4.24) to concentrate the stream of primary electrons into a powerful beam

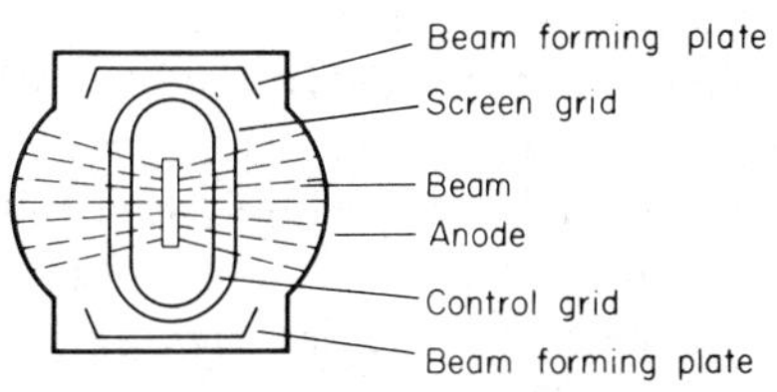

figure 4.24 Plan view of beam tetrode

that drives secondary electrons back to the anode. The beaming effect increases the maximum available anode current and beam tetrodes are therefore used in amplifiers carrying signals at high power levels. Method (3) produces a valve called a *pentode,* for there are now five electrodes. Pentodes are widely used in all kinds of valve amplifier. The suppressor grid, which is held at the same potential as the cathode, prevents the effect of the screen grid being felt at the anode and so modifies the I_A/V_A curves to those shown in figure 4.25. As can be seen, the general shape is similar to the

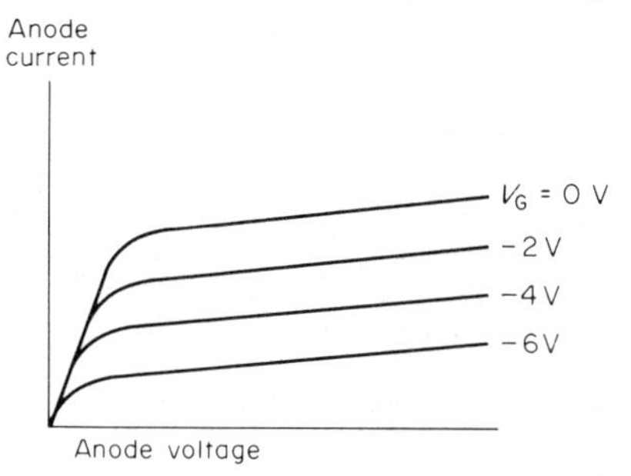

figure 4.25 Pentode I_A/V_A curves

corresponding curves of the FET. The FET is thus a device having the same number of electrodes as a triode but a performance similar to that of a pentode. The significance of this shape of curve is examined in chapter 6.

Other multi-electrode valves include the hexode and heptode, which have six and seven electrodes respectively. The additional electrodes are extra grids used for the introduction of additional signals in mixing circuits and similar applications that will be examined in due course.

4.8. The Cathode Ray Tube

The cathode ray tube is a special kind of multi-electrode valve that is capable of providing a visual display of voltage and current waveforms and is also the basis of the television picture tube and radar display tube. The basic principle is that a beam of high energy electrons, obtained in the same way as in the thermionic valve, is allowed to strike a specially coated *screen.* The screen material absorbs the kinetic energy of the electrons and re-radiates it in the form of light energy. Thus, the screen glows where it is struck by the electron beam. This property of the screen material is called *fluorescence.*

A cross-section of a typical electrostatic CRT is shown in figure 4.26. The cathode, heater and grid assembly is collectively called the *electron gun.* The grid that controls the density of the beam, and thus the brightness, is called a *modulator* in the CRT. Focussing of the electrons into a narrow beam to give a fine spot on the screen is achieved by a system of anodes that set up a suitable electric field pattern. The anode system is at a very high positive voltage (several thousand volts) with respect to the cathode so that the electron beam is attracted to it just as in the other valves; in the CRT, however, the beam passes through the anodes and continues via a deflecting system to the screen. The deflecting system consists of two sets of plates set

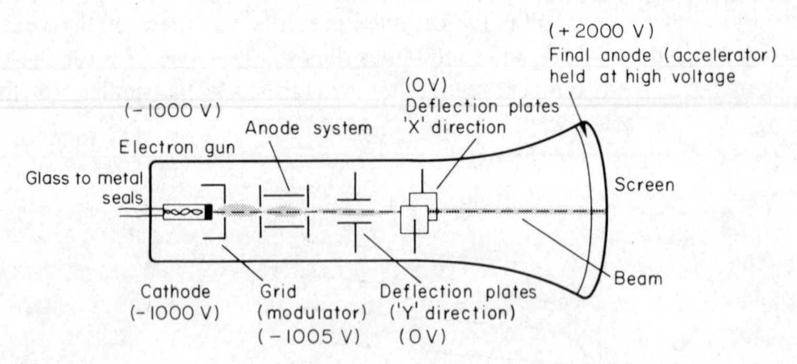

figure 4.26 Cross-section of electrostatic cathode-ray tube (bracketed figures indicate typical voltages)

at right angles to each other. Looking at the screen from the front, one set of plates deflects in the vertical or 'Y' direction, the other in the horizontal or 'X' direction. In both directions, deflection is obtained by applying a p.d. across the pair of plates so that, as the beam passes through, the electrons are attracted to the more positive of the two plates. The electrons are travelling too fast to actually hit the deflecting plates and the resultant effect is a movement of the spot in the 'Y' or 'X' direction on the screen. Once through the deflecting system the electrons are further accelerated by a post-deflection accelerating anode to provide a maximum energy collision at the screen. This anode often takes the form of a conductive ring around the largest periphery of the tube.

To obtain a larger deflection at the screen and thus a bigger display, electrostatic deflection systems such as the one described are replaced by electromagnetic systems. The movement of electrons down the tube is an electric current and, as with all electric currents, a magnetic field is set up as shown in figure 4.27. If an additional magnetic field is set up by external deflecting coils attached to the neck of the tube, interaction between the fields causes the beam to move as shown in the figure. Electromagnetic deflection is used in television and radar tubes where a large display is important.

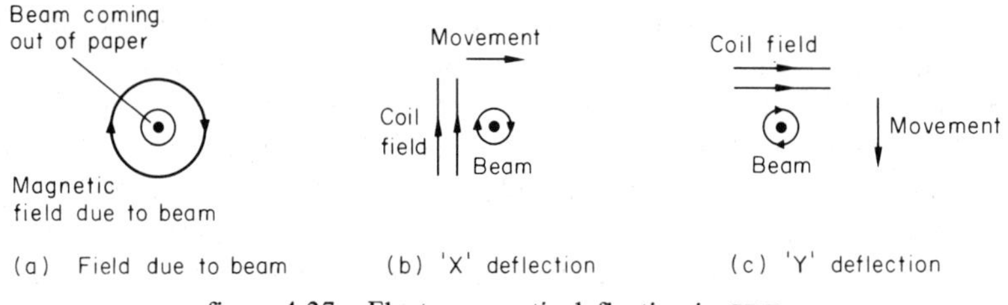

figure 4.27 Electromagnetic deflection in CRT

4.9. High-frequency and Microwave Devices

Up to about 20 MHz the normal range of diodes and multigrid valves operate satisfactorily although care must be taken to avoid unwanted feedback etc. from interelectrode capacitances at these frequencies. Special high frequency devices known as klystrons and magnetrons are used beyond this up to the ultra high frequency region (300 to 3000 MHz). Valves for use in the UHF region must be specially constructed to have very small anode-cathode separation, because the *transit time* necessary for the electron to pass from one electrode to the other at these frequencies becomes comparable with the duration of one cycle of signal frequency.

The klystron makes use of what is known as a *resonant cavity*. Probably the easiest way of understanding the basic principle of a resonant cavity is by applying the basic theory of transmission lines. Transmission lines are more fully considered in chapter 10; at the moment it is sufficient to say that a piece of conductor bent into U formation as shown in figure 4.28a behaves as

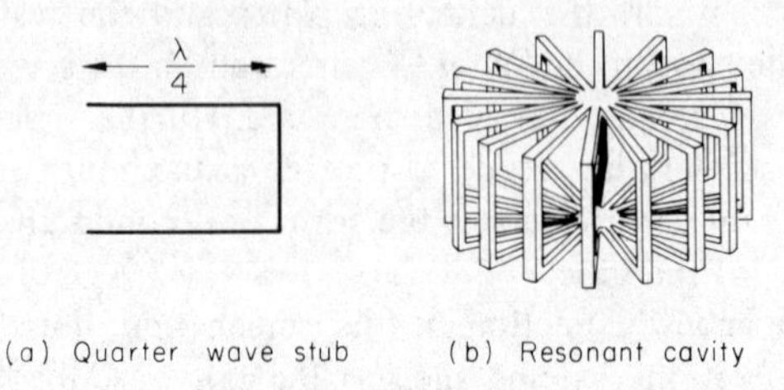

figure 4.28 Elements of the klystron

a *tuned circuit* (that is, the parallel combination of inductor and capacitor examined in section 4.3) to a signal having a wavelength equal to four times the length of one side of the U. Such a piece of transmission line is called a *quarter wave stub*. If a number of these are joined together to form a cylindrical construction as shown in figure 4.28b, the resultant shape behaves as a tuned circuit having a very sharp response curve. The advantage of such a construction is that conventional capacitors and inductors are not required; at very high frequencies individual reactive passive components are inconveniently small and would be impracticable. The resonant frequency of such a cavity is determined by the dimensions and these may be varied within a small range by inserting a movable plate to alter the internal capacitance of the device. As was mentioned earlier, and is considered in more detail in a subsequent chapter on oscillators (chapter 7), a tuned circuit is a natural oscillator. The oscillatory properties of a resonant cavity are used in the *klystron*.

In the klystron a beam of electrons is alternately speeded up and slowed down so that the beam consists of a series of *bunches* of electrons. The bunching effect of the electrons is increased by the presence of a resonant cavity and sets up alternating voltages and fields within the cavity which produce an energy output greater than the input. Thus the klystron amplifies. One form of multicavity klystron is shown in figure 4.29. The signal input is applied through a suitable transmission line (chapter 10) to the first cavity and it is here that the bunching action begins. Oscillations of increasing amplitude are set up in the following cavities as the beam is drawn to the collector plate and increased energy is drawn from the final cavity along the output transmission line. A simplified diagram of a *reflex klystron* is shown in

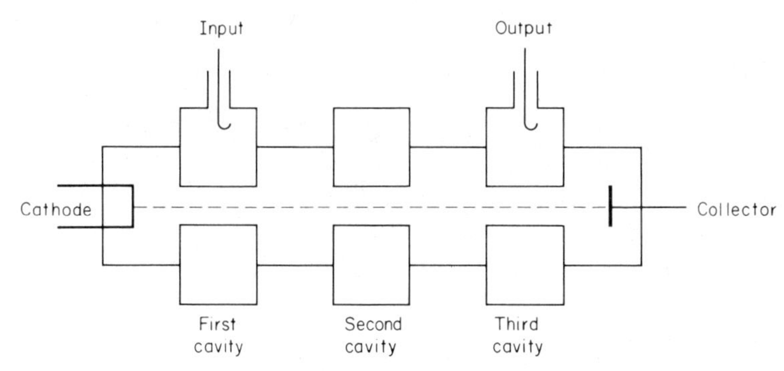

figure 4.29 Multicavity klystron

figure 4.30. Here the beam is drawn into the cavity, which is self oscillatory, and continues through the cavity grids to the end plate. In this version this plate (the repeller) is held negative and the beam returns to the cavity. If the dimensions of the path and cavity are correct, the return of the beam increases the oscillation of the cavity (*positive feedback*—see chapter 7) and an energy output is available as shown. The reflex klystron is thus an oscillator and requires no r.f. input. High power klystrons are used in some TV transmitting systems, radar transmitters and in certain special purpose communication systems.

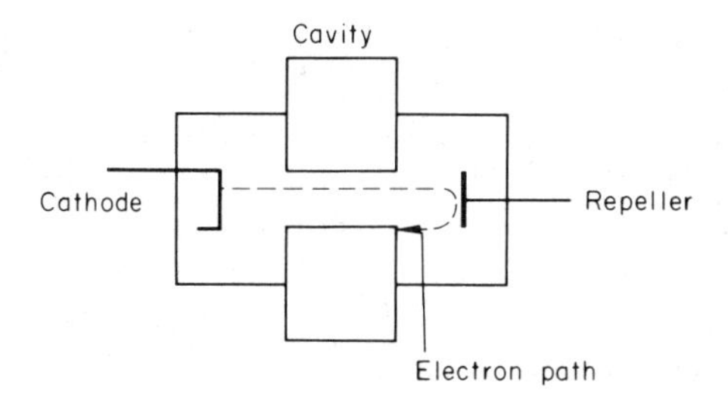

figure 4.30 Reflex klystron

Another special device used at microwave frequencies is the *magnetron,* a simplified version of which is shown in figure 4.31. Basically, it consists of a cylindrical block that has a large hole drilled axially in the centre and several smaller holes drilled parallel to the centre hole and situated between the centre and the block circumference. A heated cathode is placed in the middle of the centre hole. A strong magnetic field is applied with lines of force lying parallel to the axis of the block. If the cathode is made negative with respect to the outer block electrons emitted from the cathode will travel towards the block which behaves as an anode.

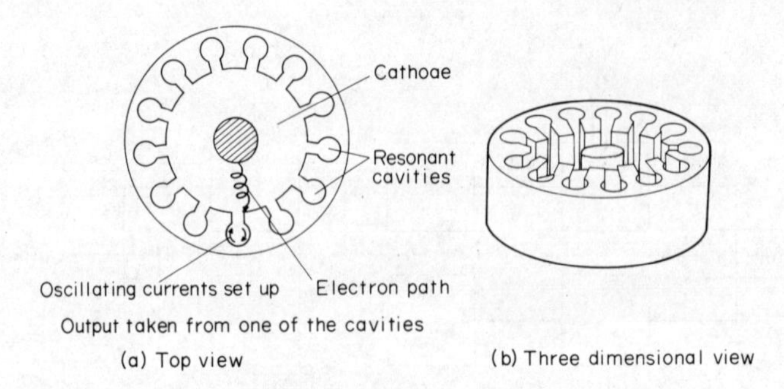

figure 4.31 Magnetron

However, the electrons do not move directly along a radius but because of the influence of the magnetic field they travel along a spiral path as illustrated in the figure. If the cathode is pulsed the spiral paths rotate much in the same way as the spokes of a wheel rotate and the electrons sweep past the minor holes. Each minor hole behaves as a resonant cavity and r.f. oscillations of voltage and current are induced in each one. R.F. energy is picked off one of the cavities by the connection of a suitable transmission line. The device is used to produce high-power pulses (in the region of 50 kW) for use in radar and similar systems.

The resonant cavity principle is used in a solid state microwave device known as a Gunn oscillator. In this component an electric field is applied to a material called *gallium arsenide,* which has an atomic structure and characteristics such that the electrons acquire different drift velocities depending upon the strength of the field. The combination of slow and fast moving electrons results in a bunching effect not unlike that produced in the klystron discussed earlier. If this bunched beam is transferred to a resonant cavity oscillations may be set up within the cavity and the device can be used as a high frequency oscillator. Frequencies up to 40 GHz are not uncommon with this component.

Other high frequency devices include *travelling wave tubes,* in which the d.c. energy given to an electron beam is extracted as a.c. energy by suitable coils mounted along the tube; and *masers,* in which certain materials are caused to radiate by stimulation using a low energy source.

4.10. Thyratrons and Thyristors

Thyratrons and thyristors are electronic components that have the common characteristic of being switched into conduction by a control grid or gate, which then loses control. Switching off either device may then be accomplished only by lowering the anode-cathode voltage. The symbols for these components and a typical IV curve are shown in figure 4.32. As indicated, the thyratron is a gas filled triode valve and the thyristor is the solid state equivalent. Both devices are widely used in power switching circuits, the advantage being that a short duration low-power pulse is able to control a high-power circuit. When used in a.c. power circuits the average

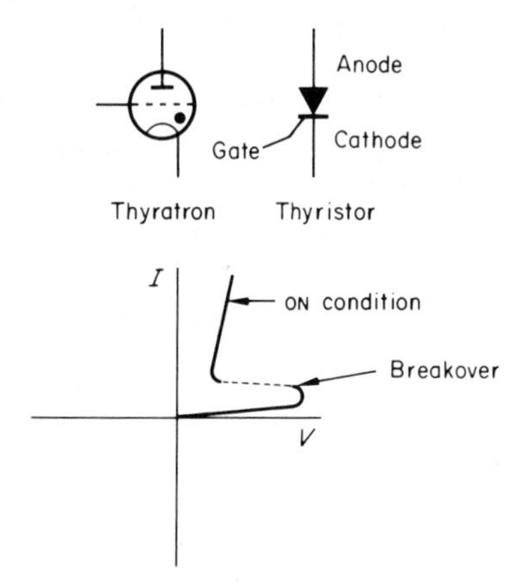

figure 4.32 Thyratrons and thyristors

power delivered to the load may be continuously varied by adjusting the point on each positive half-cycle at which the device is triggered into conduction. The device switches off during the cycle when the anode-cathode voltage falls below the extinction value. This method of control is very much better than using variable resistors, which are wasteful of energy, which they dissipate to the surroundings in the form of heat.

4.11. Photoelectric Devices

As was stated earlier in the chapter, basic atomic theory states that electrons are bound to a particular atom in orbits or shells, which are located at a distance from the atom nucleus determined by the electron energy. Given sufficient energy, electrons may leave the atom and become free. In thermionic valves this energy is provided by heat from the filament, electrons receiving sufficient energy being then able to leave the cathode altogether. Photoelectric devices use light as the energy source. They fall naturally into three categories: *photo-emissive, photoconductive* and *photo-voltaic.*

Photo-emissive devices are those in which energy from incident light radiation causes electron emission. The amount of the emission depends not only on the level of incident light but also on the materials used. Common materials used are: caesium-antimony, which responds best to light at the violet end of the light spectrum; calcium sulphide, which has maximum response to red light; and caesium silver, which is most responsive to infra-red radiation. The reason the colour is important is that the energy carried by any electromagnetic radiation is dependent to some extent on frequency, and therefore on the wavelength of the radiation. Different materials require a different amount of energy for the release of electrons (called the *work function*) and thus respond more to certain wavelengths than to others. The current flow in photo-emissive cells may be increased by using a gasfilled device, in which initial electrons cause ionisation of the gas and subsequent release of many more electrons; or by using a *photomultiplier* technique, in which primary electrons released by light are directed on to a target that reflects the beam and adds to it by *secondary emission* (due to the absorption of energy by the secondary emissive surface). Such a process may be continued to give a very large output current (see figure 4.33).

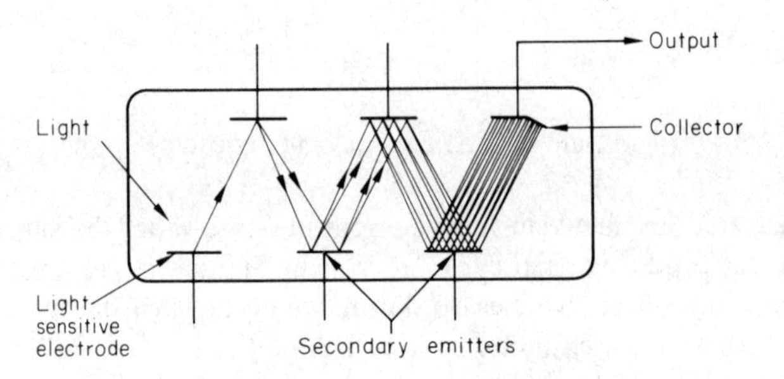

figure 4.33 Photomultiplier

Photoconductive cells contain a material having a work function such that electrons are not fully released but are nevertheless freed from their molecular bonds, to become available for charge carrying (that is, to form current) and thus the conductivity of the material increases. This effect is used in light meters in photography and is also the basis of the photodiode and photo-transistor. The saturation (reverse) current of a PN diode is sometimes light sensitive, so that if the diode is mounted in a transparent container and reverse biased it may be used as a light controlled switch. The photo-transistor employs the same principle and in addition amplifies the light dependent current. If this effect is not required solid state diodes and transistors are mounted in opaque containers.

Certain semiconductor materials (for example, silicon, boron and selenium) when arranged correctly set up an e.m.f. on exposure to light radiation. This effect is called the *photovoltaic* effect and is used in light meters that do not require any additional power source.

4.12. Integrated Circuits

Until a few years ago all electronic circuits were made up of individual (or *discrete*) components connected together by wires. These connections were either separate conductive leads, or else were copper strips arranged on a printed-circuit board (that is, a copper-coated insulating board, such as bakelite, from which all surplus copper has been etched leaving only the required circuit pattern). With the rapid development of solid state devices during the early 1960s, however, it became possible to manufacture more and more components in a single piece (or *chip*) of semiconductor material. At the same time development proceeded on smaller and smaller discrete components to be used with the new varieties of *film circuits* then becoming available. It is now possible to think in terms of 10 000 components per cubic centimetre, using one or more of the various *integrated circuit* techniques. A single integrated circuit can already contain almost a complete radio receiver, four or five a complete television receiver. Computers, which once required several cubicles two or more metres high are now contained in a single desk unit.

Integrated circuits may be divided into two kinds, *film* circuits and *monolithic* circuits. A film circuit is made by printing patterns of conductive (or resistive) material on to a ceramic base, called a *substrate* and firing the mixture in suitable ovens to a hard finish. Using this process it is possible to 'print' resistors, capacitors and to some extent inductors and their connecting leads in one manufacturing process. Microminiature active components or monolithic chips, as described below, may then be added to the finished film circuit using *compression bonding* (exerting controlled pressure at an appropriate surrounding temperature) or some similar technique.

Monolithic circuits are composed of a single chip of semiconductor containing diodes, transistors and passive components (except most inductors) which are manufacturered together in the same process. It is now accepted throughout the electronics industry that to take a piece of P-type material, dope a layer of controlled thickness of N-type material, then repeat the process with P-type material in order to make one single PNP *planar* transistor (as shown in figure 4.34) is a time-wasting process when it is just as convenient to make several hundred in the same chip. In fact, this process is now used to make discrete planar transistors, the chip being divided by diamond cutters after manufacture. PN diodes, PNP and NPN bipolar transistors and FET's may all be made using this planar process, the impurity

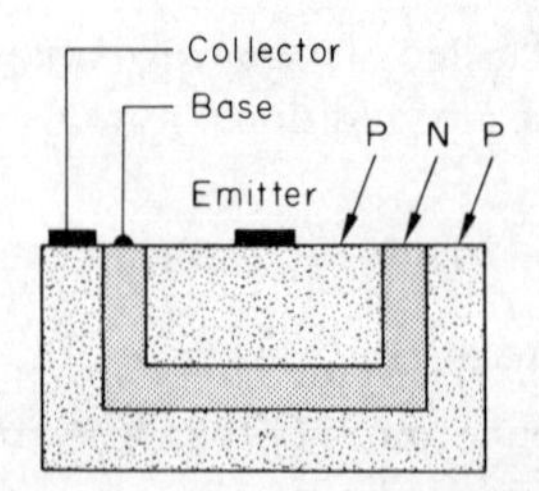

figure 4.34 PNP planar transistor

required to produce N-type or P-type properties being introduced by a *diffusion process* (involving passing a hot gas containing the additive over the surface for a controlled period of time). Resistors in a monolithic chip are a layer of either P or N material suitably doped to give the required level of resisitivity. Capacitors consist of two conductive layers separated by an appropriate oxide layer to act as dielectric. Spiralling a conductive layer to increase inductance is a possibility but most inductors are added externally, the problem here being the core material. Care should be taken by the mechanic when attempting to understand the equivalent circuit of an integrated circuit supplied by a manufacturer, because it is common practice to use the resistive or capacitive properties of (say) a transistor instead of using an actual resistor or capacitor, the reason being that it is often easier and cheaper to mass-produce hundreds of transistors simultaneously and then print the interconnecting pattern of conductive leads, rather than to halt and re-arrange the process in order to manufacture resistors and capacitors. A small part of a typical integrated circuit is shown in figure 4.35.

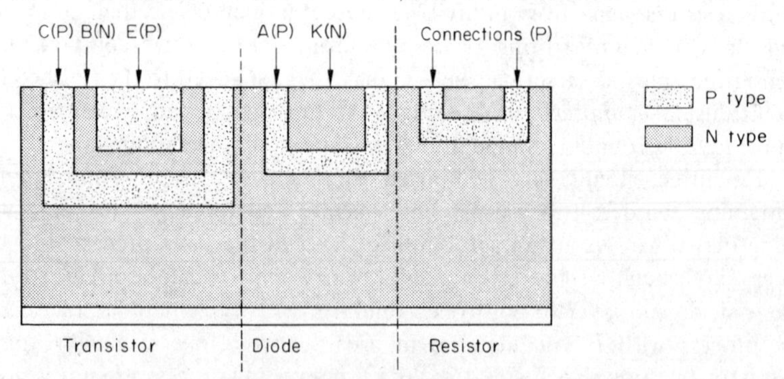

figure 4.35 Integrated circuit components

5 Power supplies

(The reader is advised to read chapter 6 on active device operation before attempting article 5.3 in this chapter.)

Power supplies for electronic systems fall into three types: battery derived supplies; mains derived supplies; and supplies derived from local generators. Sometimes more than one type or a combination of them is used, depending upon the nature of the system. Small portable systems (for example, transistorised radios, televisions and certain telecommunications equipment) invariably use batteries; larger fixed systems derive their power from the national electricity supply; and large portable systems, in air or seaborne vehicles for example, carry their own generating machinery. Systems such as those used in satellites or interplanetary vehicles normally use battery supplies, the initial energy being obtained via suitable transducers from sunlight.

Batteries fall into two types, primary and secondary batteries. Primary batteries cannot be recharged once the energy is used and must therefore be replaced. Secondary batteries can be recharged by a suitable source of electrical energy and do not have to be replaced except after very long periods of use. Systems using batteries require no special circuitry and they require less components. The obvious disadvantage is that the battery must eventually be renewed.

Generating machinery has the disadvantage that maintenance represents a continuing commitment. In addition, any *prime mover* used to drive the generator requires fuel and adds yet further maintenance. Where mains electricity is not available, however, a generator of some kind is usually essential. Aeroplane and ship systems invariably employ local generators.

Probably the most commonly used supply, certainly in land based systems, is one derived from the mains electricity supply. Since the national supply is usually alternating current for ease of transmission and distribution, additional circuitry is required to convert or *rectify* the a.c. supply to the direct current normally required by electronic circuits.

5.1. Rectification

Rectification is the process of converting alternating current (which flows in two directions within an electric circuit) into direct current (which flows in one direction). To do this, a device is required which has a very low resistance in one direction and a very high resistance in the other direction. A diode meets these requirements.

Figure 5.1a shows the simplest form of diode rectifier circuit consisting of an a.c. supply fed to a diode and load resistor in series. When the a.c. supply

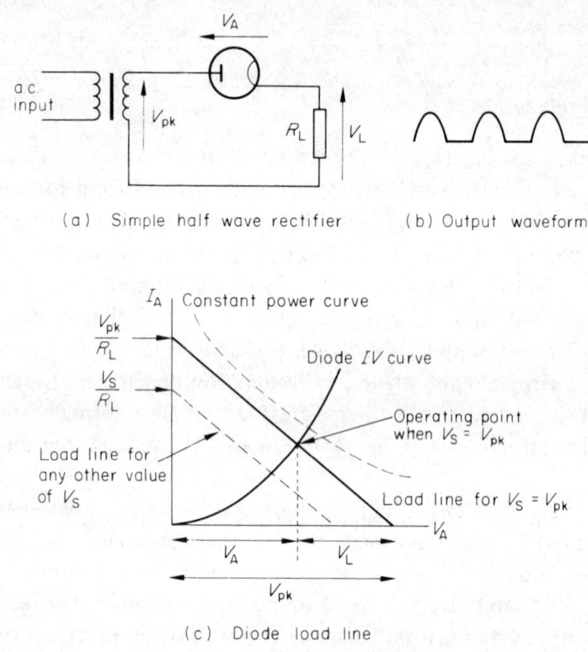

figure 5.1 Simple diode

makes the anode positive with respect to the cathode a large current flows, its value being determined by the supply voltage and the combined forward resistance of the diode and load resistance. When the anode becomes negative with respect to the cathode no current flows if the diode is a valve, but a very small current flows if the diode is a semiconductor. Thus, current flows in a series of unidirectional pulses as shown in figure 5.1b. This process is called half-wave rectification because the large current flows on only one half of each input cycle.

In the circuit of figure 5.1a the supply voltage at any one time equals the sum of the diode voltage and the load voltage. Using the symbols shown on the diagram and considering the instant when the supply voltage is at a peak value

$$V_{pk} = V_A + V_L$$

As the load voltage is the product of the load resistance and the diode current

$$V_L = I_A R_L$$

so that the expression becomes

$$V_{pk} = V_A + I_A R_L.$$

If this equation is rearranged to give the diode current in terms of the other quantities, then

$$I_A = -\frac{V_A}{R_L} + \frac{V_{pk}}{R_L}$$

and this equation indicates how anode current and anode voltage vary for a given value of supply voltage and load resistor. If a graph of I_A against V_A is plotted from this equation as shown in figure 5.1c, it turns out to be a straight line cutting the I_A axis (where $V_A = 0$) at $I_A = V_{pk}/R_L$ and cutting the V_A axis (where $I_A = 0$) at $V_A = V_{pk}$. This line is called the *load line* for the diode operating in a half-wave circuit. If the diode I_A/V_A curve is drawn on the same graph the position of the load line for a given value of supply voltage relative to the diode curve gives the value of the diode voltage and the load voltage. The diode curve shows how anode voltage and anode current are related, the load line shows how load voltage and anode current are related with a load in circuit. The intersection of the two curves gives the operating point of the valve as this is the only point that satisfies both the I_A/V_A requirements of the diode and the I_A/V_L requirements of the load. The values of the respective voltages are shown in the figure. When the supply voltage is at any other value, say V_S, the load line moves down the diode curve as indicated by the dotted line. The new load line is parallel to that for peak voltage, the points of intersection being V_S/R_L on the I_A axis, and V_S on the V_A axis.

Two important characteristics of the diode rectifier which must be borne in mind when selecting a particular diode are *power dissipation* and *peak inverse voltage* (PIV).

Diode power dissipation is determined by anode voltage and anode current. For a particular diode a maximum allowable dissipation is stated by the manufacturer. Exceeding this figure damages the diode. For a particular power level W, for the diode

$$I_A V_A = W,$$

so that

$$I_A = W/V_A$$

and if this graph is plotted it gives the constant power dissipation curve shown in figure 5.1c. The curve gives the value of anode current for any given value of anode voltage to give a power dissipation of W watts. Maximum diode power dissipation occurs, obviously, when V_A and I_A are at a maximum. This in turn occurs when the supply voltage is at peak value. The operating point shown in the figure must not then be further up the I_A/V_A curve than the point of intersection between the I_A/V_A curve and the curve for maximum allowable power. If it is, the diode power will exceed the maximum safe limit. The graphical approach used in this article is extremely useful and should be carefully noted. It is used again in chapter 6 when amplifiers are discussed.

Diode peak inverse voltage is the maximum safe value of voltage that can be applied in the reverse direction (anode negative) without risk of breakdown. In circuits using a *reservoir capacitor* (as about to be described) the maximum voltage applied in the reverse direction can be as much as twice the peak value of the supply; it is therefore essential to consider the PIV of the diode carefully before connecting the diode into a circuit.

Full wave rectification is illustrated in figure 5.2. In these circuits more than one diode is used and their arrangement is such that both halves of each a.c. input cycle are used to give an output current wave as shown in figure 5.2b. In the full-wave circuit of figure 5.2a two rectifiers are used. When the anode of one is positive with respect to its cathode, the anode of the other is

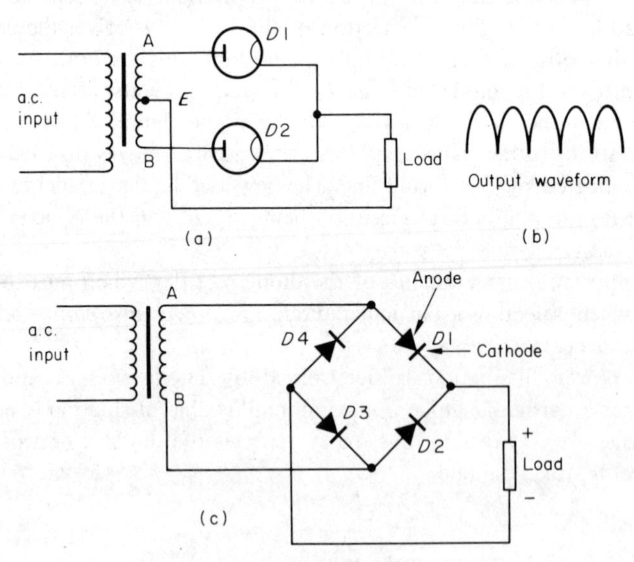

figure 5.2 Full-wave rectifier circuits

negative with respect to its cathode. At each reversal of input polarity the anode-cathode voltages change their polarity such that one diode of the two conducts whilst the other is cut off. In the following half cycle the conducting diode cuts off and the other conducts. Thus, a unidirectional current in pulse form is obtained as shown in figure 5.2b. In detail, when the top of the transformer secondary (point A) is positive with respect to the earthed centre tap (point E) the bottom of the secondary (point B) is negative with respect to the centre tap. Thus diode *D*1 has its anode positive and conducts, diode *D*2 has its anode negative and is cut off. When the input a.c. wave reverses polarity point A becomes negative with respect to point E and *D*1 cuts off, point B becomes positive with respect to point E and *D*2 conducts. If the diodes were solid-state they would not, of course, completely cut off in the reverse direction but a very small saturation current would flow. The output current would then be reduced slightly on each half wave (by equal amounts if the semiconductor diodes were matched in the reverse direction).

One disadvantage of the circuit of figure 5.2a is the need for a centre-tapped input transformer. The circuit of figure 5.2c called a full-wave *bridge rectifier* does not have this disadvantage. It does however require two more diodes. In the bridge circuit shown, when point A is positive with respect to point B the anodes of *D*1 and *D*3 are positive with respect to their respective cathodes and current flows from A via *D*1 to the load then to *D*3 and back to point B. Diodes *D*2 and *D*4 are reverse biased in this input half-cycle. When B becomes positive with respect to A, diodes *D*2 and *D*4 are forward biased and current flows from B to *D*2 then to the load then to *D*4 and back to A. In this half-cycle diodes *D*1 and *D*3 are reverse biased. In both half cycles the current flows in the same direction through the load giving an input voltage across the load resistor of the polarity shown. The waveform is the same as for the previous circuit and is shown in figure 5.2b.

Both half and full wave rectified outputs are direct current in the sense that current flows in only one direction. However, the simple circuits shown give a pulsating output and in the majority of applications this varying nature of the current is undesirable. Additional components are thus required to *smooth* and *filter* the pulsating output.

5.2. Smoothing and Filter Circuits

Figure 5.3a shows a capacitor-input smoothing and filter circuit made up of a capacitor *C*1, called the *reservoir capacitor,* and an inductor-capacitor filter made up of components *L*1 and *C*2. To explain this circuit, the effects of *C*1 and of *L*1 and *C*2 will be examined separately.

If a capacitor is connected across the output of the circuit of figure 5.1a (the half wave circuit) and the load is removed, the output waveform becomes that shown in figure 5.3b. Similarly the output of the full wave circuits shown in figures 5.2a and 5.2c would become that shown in figure 5.3c. As has been explained, a capacitor cannot follow rapid voltage changes because time is required for the level of charge on the plates to adjust to new

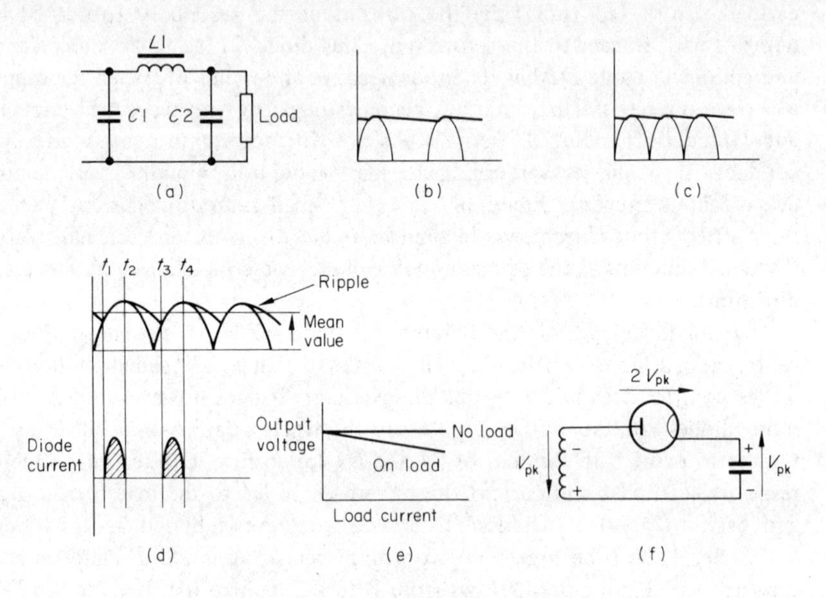

figure 5.3 Capacitor-input filter circuits

conditions. The capacitor thus charges to the peak value of the output voltage, dropping slightly when the output voltage drops below this peak value. The drop occurs due to capacitor discharge via the diode reverse resistance and is greater with solid state diodes due to their lower reverse resistances. With a load connected, the discharge path resistance drops even further and the capacitor voltage droop is further increased as shown in figure 5.3d. During period t_1 to t_2 in this diagram the capacitor is charging as the output voltage rises. Diode current flows during this period as shown in he lower graph of the figure. During period t_2 to t_3, the output voltage falls below the peak value and the capacitor discharges through the load resistance and diode reverse resistance as previously explained. The capacitor voltage continues to fall until the rising output voltage is equal to it at t_3. The charging process begins again and continues during period t_3 to t_4, then cuts off again until the next cycle repeats exactly the same sequence of events.

The amount of 'droop' depends to a large extent on the capacitance, the droop being small for large capacitance values. Care must be taken not to make the capacitor too high a value, however, because this reduces the period t_1 to t_2 (or t_3 to t_4) during which diode current flows and the current must therefore be higher in order to fully charge the capacitor during the shorter period. Too small a value of capacitance will of course greatly increase the droop.

The output voltage waveform can now be considered to be made up of two components, a d.c. level indicated by the 'mean value' arrow in the figure and an alternating component called the *ripple.* Examination of the diagrams shows that the ripple frequency is equal to the supply frequency for half-wave rectification and to twice the supply frequency for a full-wave rectified output.

If now $L1$ and $C2$ are connected between the reservoir capacitor $C1$ and the load, the ripple amplitude may be very much reduced without much affecting the d.c. output level. The inductor $L1$ is chosen so that it offers a high impedance to ripple but low resistance to the d.c. component. The output capacitor $C2$ is chosen so that it offers a low reactance to ripple but high resistance to d.c. Thus the output voltage will be made up of a slightly reduced d.c. level together with a much reduced a.c. ripple. The loss in the direct voltage output depends on load current and increases with increasing load current as shown in figure 5.3e. This curve is called the *voltage regulation curve* of the supply. Occasionally the inductor is replaced by a resistor in low grade power supplies for price economy. Such a circuit is not as effective, however, because the replacement resistor offers an equal resistance to d.c. and a.c., so that increasing its value reduces ripple but also reduces the output thereby worsening its regulation, and reducing its value improves the regulation but increases the ripple. A poorly filtered supply will cause 50 Hz or 100 Hz 'hum' in a loudspeaker if the power supply is feeding a radio or television system.

Circuits employing reservoir capacitors have two disadvantages: firstly, that high current pulses flow in the rectifier during the charging period as already explained; secondly, that during non-conducting periods the maximum voltage applied to the rectifier can be as much as twice the peak voltage of the supply. The reason for this is that the capacitor charges to the peak of the supply, and this voltage is then in series with the a.c. supply across the diode. This is shown in figure 5.3f.

If the reservoir capacitor is removed ($L1$ and $C2$ remaining) the circuit is called a choke-input filter. The half or full wave rectified output can again be considered to be made up of a d.c. level and an alternating ripple component, the ripple this time being somewhat larger than it was when the reservoir capacitor was present. As before the ripple is reduced by the filter circuit. One disadvantage of this filter circuit is that at low load currents the output voltage rises rather more sharply than with the capacitor-input circuit. The regulation curve is shown in figure 5.4b and the circuit in figure 5.4a. Use of variable inductance or *swinging chokes* having a higher inductance at low

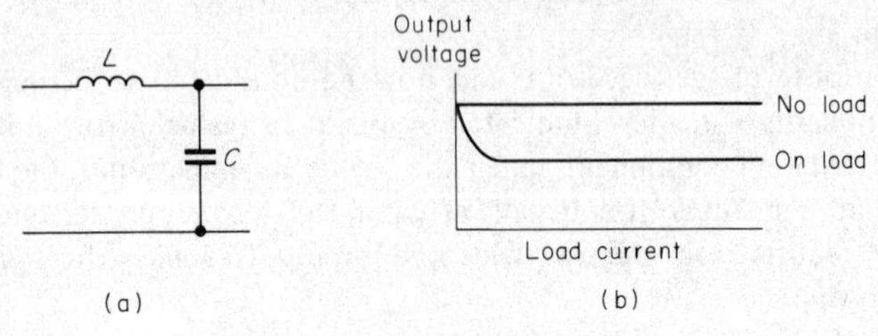

figure 5.4 The choke-input filter

current values helps to compensate for the sharp rise. Another method, also employed with capacitor-input filters, is to prevent the load current falling below a pre-set minimum value by using a permanently connected *bleeder resistor* across the output. A complete full-wave circuit using such a resistor is shown in figure 5.5.

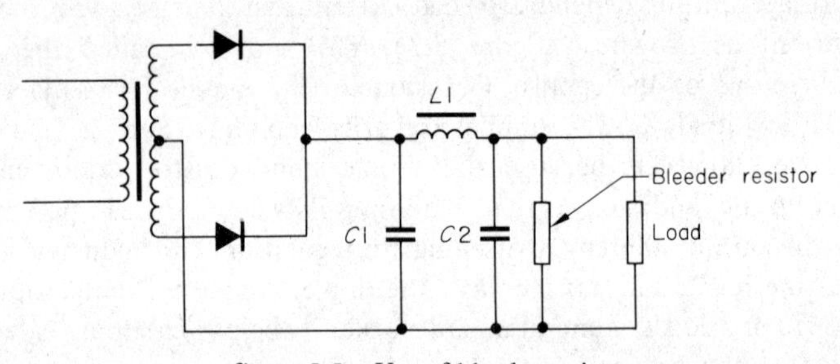

figure 5.5 Use of bleeder resistor

5.3. Stabilisation

The regulation curves of the circuits already discussed may be improved by stabilisation. Voltage stabilisers, including the cold-cathode gas filled diode valve and the solid-state Zener diode, were discussed in chapter 4. Both these diodes maintain a fairly constant voltage over a wide range of current values.

A simple voltage stabiliser circuit using either a cold-cathode valve or a Zener diode is shown in figure 5.6a. Resistor R_L is the load resistor and is shunted by the stabilising diode. The unstabilised input from a power supply circuit (which could be any of those already considered) is fed to the

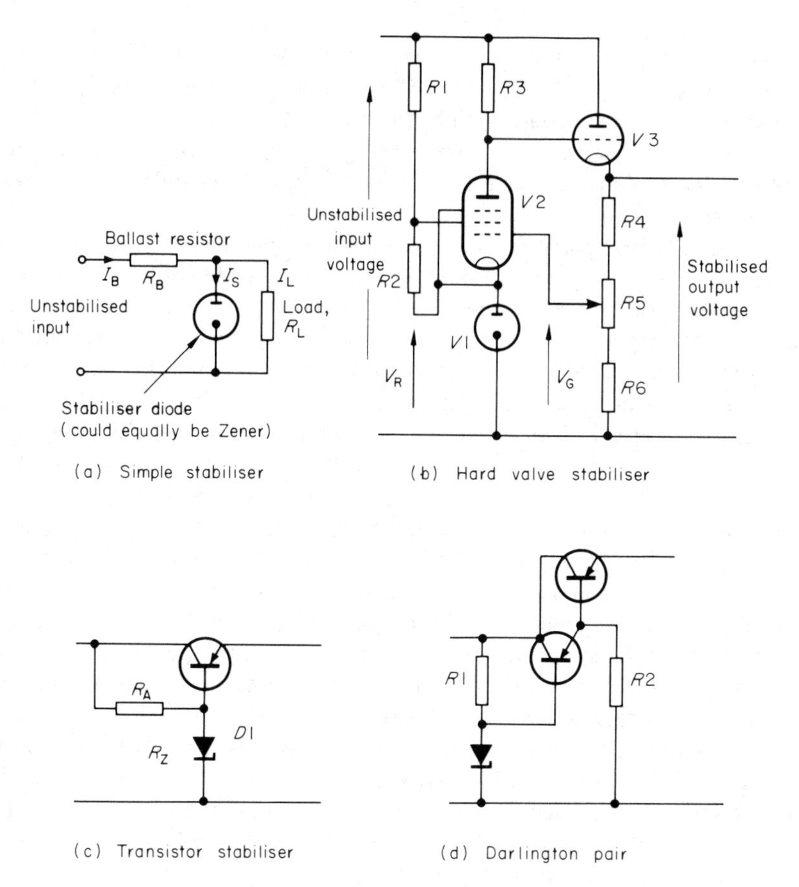

figure 5.6 Voltage stabilisation

load/stabiliser combination via resistor R_B, which is called the *ballast resistor.* The circuit stabilises both against input changes and against load changes. For a constant input voltage the voltage drop across R_B and the output voltage remain fairly constant provided that the stabiliser current I_S lies within the range of values given by the manufacturer. If the load current I_L rises and I_S falls, the sum, which is the ballast current I_B, remains constant. Similarly, if I_L falls, I_S rises and again I_B remains the same. As long as I_B is constant the voltage drop across R_B and thus the output voltage remains constant. If the load current is such a value that the remainder of I_B (which is the stabiliser current) does not lie between the limiting values the stabiliser no longer holds the voltage constant. At low values of load current, when the stabiliser current is high, care must be taken not to damage the stabiliser. A similar process takes place for input voltage variations. Here, the ballast resistor voltage drop adjusts to take the difference between the changing input and the constant output voltage. Again, the output voltage remains constant only if the stabiliser current lies between the given limiting values. If the input voltage changes to a value such that this is not so, then the output voltage will change accordingly.

A vacuum valve stabiliser is shown in figure 5.6b. Valve $V1$ is a stabiliser diode as used in the previous circuit and sets the cathode voltage of valve $V2$. The control grid voltage is determined by variable resistor $R5$, the difference between this voltage and the stabiliser voltage being the bias voltage applied to valve $V2$. The anode voltage of $V2$ controls the bias on valve $V3$ which is in series with the load across the unstabilised supply. The output voltage is across the potential divider chain $R4$, $R5$ and $R6$ and determines voltage V_G applied to the grid of valve $V2$. At the required output $R5$ is adjusted so that $V_R = V_G$. If the output voltage rises, V_G rises and the bias of valve $V2$ changes to allow a higher anode current. Thus the anode voltage of $V2$ falls, changing the bias of series regulator valve $V3$ and increasing its resistance so that the voltage drop across $V3$ rises. This reduces the output voltage back to the desired level. Similarly, if the output voltage falls, V_G falls, valve $V2$ anode current falls and $V2$ anode voltage rises. The bias on $V3$ changes reducing its resistance and the voltage drop across it, thereby raising the output voltage back to the required level.

A transistorised series regulator is shown in figure 5.6c. The circuit is in fact an *emitter follower* which passes the constant voltage of the zener diode to the output. This circuit has the distinct advantage of current gain, which means that much higher load currents can be handled with the same type of diode as used in the circuit of figure 5.6a. Even larger currents are obtainable with the circuit shown in figure 5.6d. This circuit is called a Darlington pair, the overall current gain being the product of the gains of the two. The zener

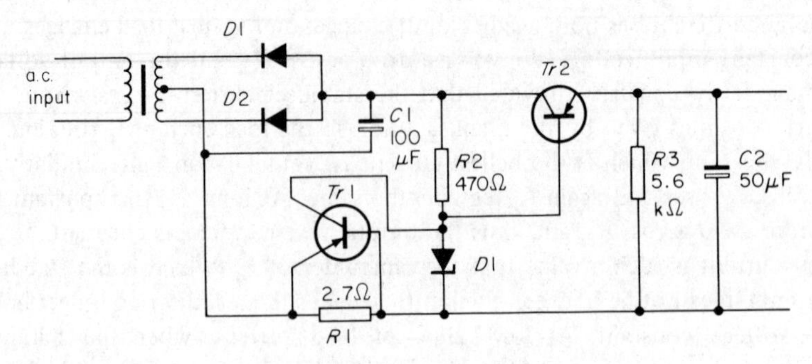

figure 5.7 Overload protection

diode operates at a very much lower value than that of the load current, which can be in the ampere range.

In an unregulated power supply when the load current rises, the output voltage falls due to the increased voltage drop across the internal resistance of the supply. In the series regulator circuit of figure 5.6c, if the load current rises, any tendency of the output voltage to fall increases the transistor forward bias (the bias on the regulator transistor is of course the difference

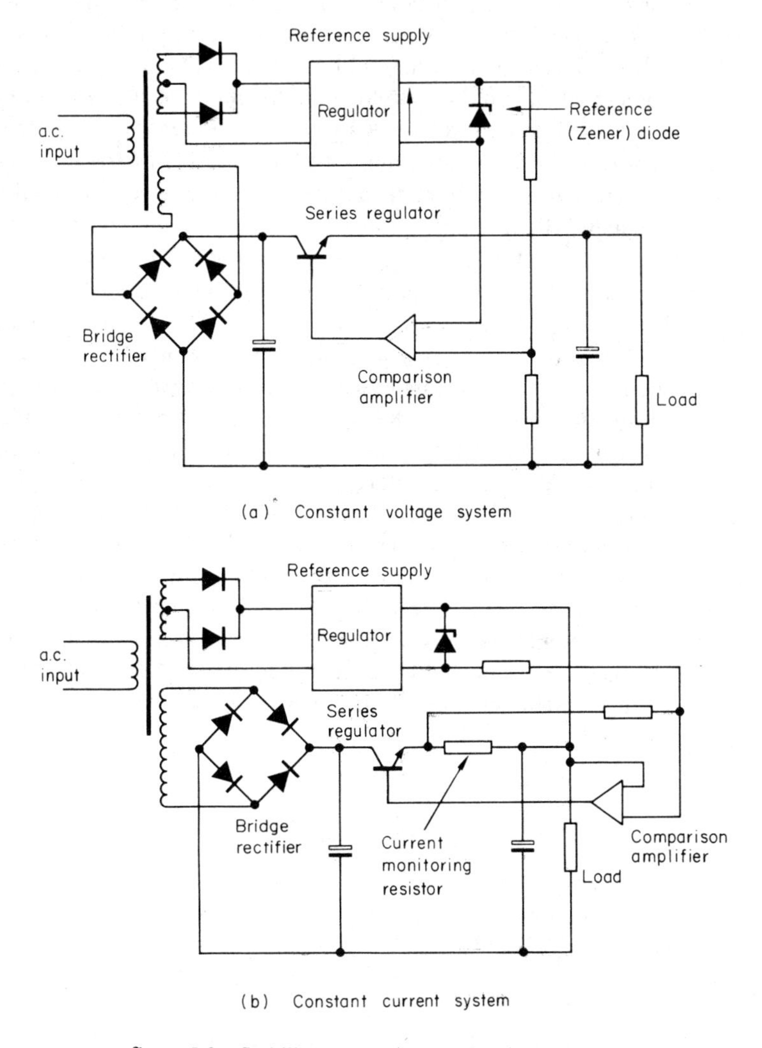

figure 5.8 **Stabilisation using operational amplifiers**

between the zener voltage and the output voltage) thereby *reducing* its effective resistance and thus the voltage drop across it, so that the output voltage is restored to the desired value. Similarly, if the load current falls, which in an unregulated supply causes a rise in output voltage, the reduced forward bias increases the regulator resistance and thus the voltage drop, thereby forcing the output voltage down to the stabilised value. The potential divider chain formed by resistor R_A and the zener diode, which sets the regulator base-earth voltage, also reduces ripple in the ratio $R_Z/(R_A + R_Z)$ where R_Z is the zener diode resistance. Since an emitter follower has no voltage gain the reduced ripple is that which appears across the output. For maximum ripple control and d.c. stability, R_A should be as high as possible.

Figure 5.7 shows a series-regulator with overload protection. Resistor $R1$ together with the load current determines the bias on overload transistor Tr1. Should the output current attempt to rise above a preset level, the overload transistor conducts via $R2$ thus protecting the regulator transistor from excessive load current. A trip switch may be incorporated in the overload circuit to disconnect the power supply output from the load in the event of an overload.

Higher stability and even better regulation may be obtained by additional regulator circuits, one such system using an *operational amplifier* (see chapter 6) is illustrated in figure 5.8. Figure 5.8a shows a constant voltage system in which the output voltage is compared using an operational amplifier with a preset reference voltage unaffected by load. The amplifier controls the bias of a series regulator transistor which controls the output by changing its own resistance in a similar manner to that in the simple circuits discussed earlier. The constant current system illustrated in figure 5.8b works in a similar way, the reference voltage being compared with the voltage drop across the current monitoring resistor shown. The difference between the two voltages controls the regulator bias and thus the output current. The blocks shown as reference regulators in both diagrams could be circuits of the form shown in figure 5.7.

6 Amplification

Chapter 4 described a number of components in which the through current between two electrodes is controlled by the voltage applied at a third electrode. These components included both types of transistor and single and multigrid valves, including triodes, tetrodes and pentodes. It was said then that any component of the type described in which one electrode voltage had more effect on the current than another electrode voltage, is capable of amplification. This chapter is concerned with detailed circuits involving active and passive components which are used for amplification.

6.1. Basic Principles: Modes of Operation

To obtain voltage amplification the current from the valve or transistor must be passed through a circuit offering impedance so that an *output voltage* is developed. The ratio of this output voltage to the voltage controlling the current is then the *gain* of the amplifier. The circuit across which the output voltage is developed, which is called the *load,* may be a resistor, a *tuned circuit,* or even another valve or transistor, depending upon the type of signal that is to be amplified. Of the three electrodes available (control grid, cathode and anode; or base, emitter and collector; or gate, source and drain), the electrode to which the load is connected is known as the output electrode, the electrode to which the signal to be amplified is connected is called the input electrode and the remaining electrode is common to both input and output circuits. Thus, there are three possible *modes* of amplifier as illustrated in figure 6.1. For the valve (triode, tetrode or pentode) there is the *common cathode, common grid* and *common anode* (or *cathode follower*); for the bipolar transistor the *common emitter, common base* and *common collector* (or *emitter follower*); and for the FET *there is the common source, common gate* and *common drain* (or *source follower*). It is found on comparison between similar modes (for example, common cathode, common

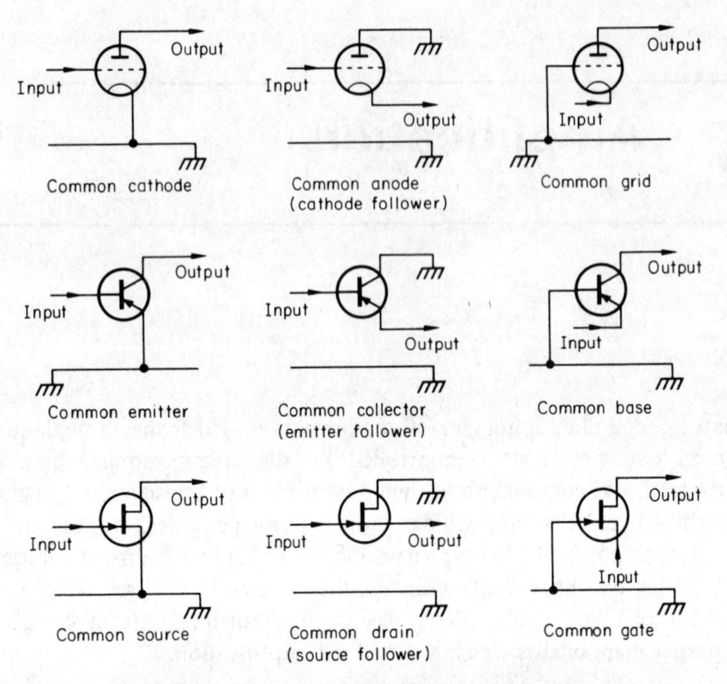

figure 6.1 Modes of operation of valves and transistors

emitter, common source) that the overall amplifier characteristics are similar; the important characteristics being input and output impedances, voltage gain and transfer characteristics. This point will be examined in more detail shortly.

Consider the simple circuit shown in figure 6.2. The circuit shows a triode valve connected in series with a resistor across a supply voltage, V_S. The supply voltage V_S, which is constant, is equal to the sum of the load voltage V_L and the valve anode-cathode voltage V_A ; that is

$$V_S = V_L + V_A$$

The load voltage V_L is equal to the product of the current through it, the anode current I_A, and the resistance of the load R_L; that is

$$V_L = I_A R_L$$

which gives the supply voltage

$$V_S = I_A R_L + V_A$$

at all values of anode current. If the anode current changes the load voltage must change and, because the supply voltage is constant, the anode voltage V_A must change. In detail, if the anode current *rises,* the load voltage *rises* and so the anode voltage must *fall.* Similarly, if the anode current *falls,* the load voltage *falls* and the anode voltage *rises.* To make the anode current rise or fall the grid-cathode voltage is altered. If the grid is made less negative, anode current rises, and if the grid is made more negative, anode current falls.

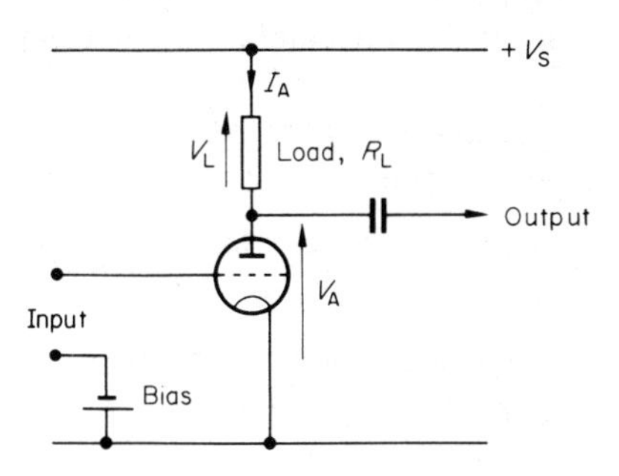

figure 6.2 Simple amplifier

Thus, altering the grid voltage changes the anode current and, because of the circuit arrangement, the anode voltage. In this simple circuit the input is applied between grid and cathode, the output is taken from between anode and cathode. The cathode is common to both input and output circuits and thus the circuit is the basis of a *common cathode* voltage amplifier. The circuit shows a triode valve but the principle applies equally to other valves and to both types of transistor circuit. As can be seen, in this mode of amplifier the output voltage rises when the input voltage falls, so the output is in antiphase with the input and there is a half cycle or 180° phase shift through the circuit.

The ratio between the change in anode voltage due to a change in grid voltage is known as the voltage *gain* and it depends upon a number of factors. These include firstly, the change in anode current produced per volt change in the grid voltage; and, secondly, the value of the load resistance. This is fairly clear from the operating equation already formulated which shows that the change in V_A is caused by change in V_L.

The ratio between change in anode current per volt change in grid-cathode voltage is determined by the actual valve used. This ratio is, in fact, called the *mutual conductance* and its symbol is g_m. Mutual conductance is one of a number of *parameters* that are used to predict valve or transistor performance. Device parameters are briefly discussed in the next article and are described in detail in appendix 3.

6.2. Valve and Transistor Parameters

Valves or transistors made to do the same job (for example, to amplify audio frequency signals or radio freqency signals) may be compared by using their *parameters.* Parameters are considered in detail in appendix 3, but a summary of the commonly used parameters is as follows.

There are three important valve parameters (see figure 6.3). These are anode resistance (r_a), mutual conductance (g_m) and amplification factor (μ).

$$\text{Anode resistance} = \frac{\text{small change in anode voltage}}{\text{small change in anode current}} \text{ (with grid voltage constant)}$$

$$\text{Mutual conductance} = \frac{\text{small change in anode current}}{\text{small change in grid voltage}} \text{ (with anode voltage constant)}$$

$$\text{Amplification factor} = \frac{\text{small change in anode voltage}}{\text{small change in grid voltage}} \text{ (with anode current constant)}$$

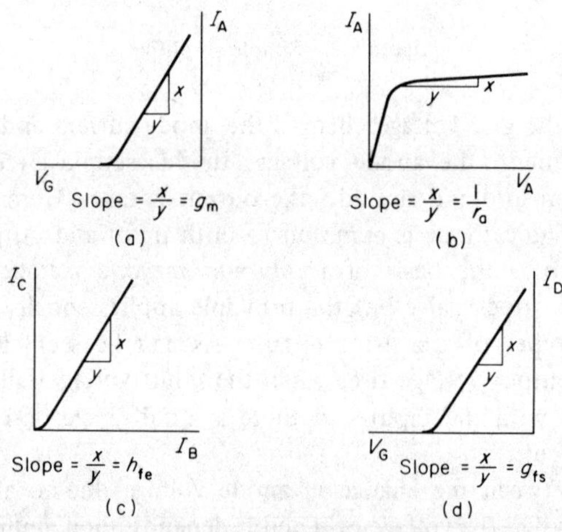

figure 6.3 Parameters

Anode resistance is a measure of the valve resistance in circuit and indicates the output impedance of a one valve amplifier. Both mutual conductance and amplification factor are measures of how well the valve amplifies, for g_m shows how the grid voltage affects anode current and μ shows the relative effect of a grid voltage change to an anode voltage change. Although in practice an amplifier containing one valve can never have the gain equal to μ, nevertheless μ may be taken as a reliable guide to the possible gain. As is shown in appendix 3, r_a, g_m and μ are related by the equation

$$\mu = r_a \times g_m$$

which is a useful relationship to remember. As an example of the calculation of the parameters consider a valve on which the following measurements were made:

Anode voltage	Anode current	Grid voltage
100 V	5 mA	−2 V
150 V	10 mA	−2 V
150 V	5 mA	−3 V

With grid voltage at −2 V, an anode voltage change of (150 − 100) V produces a current change of (10 − 5) mA so that r_a = 50/5 = 10 kΩ. With anode voltage constant at 150 V, a grid voltage change from −2 V to −3 V produces a current change of (10 − 5) mA so that $g_m = 5/1$ milliamperes/volt. Hence $\mu = r_a g_m$ is given by $\mu = 10 \times 5 = 50$. A single stage valve amplifier using this valve could have a gain of over 40 depending upon the amplifier load. (It could not have a gain of 50 because some voltage is 'lost' across the valve resistance. See article 6.3.)

Important transistor parameters are h_{fb} and h_{fe} (see figure 6.3c) which are the current gain from emitter to collector and from base to collector respectively.

$$h_{fb} = \frac{\text{small change in collector current}}{\text{small change in emitter current}}$$

$$h_{fe} = \frac{\text{small change in collector current}}{\text{small change in base current}}$$

Both these parameters are a measure of current gain in the transistor and to some extent will also determine the voltage gain of a voltage amplifier. Notice that both parameters concern small changes (that is, a.c. signal quantities). The ratios of d.c. quantities, for example:

$$\frac{\text{collector current (d.c. value)}}{\text{emitter current (d.c. value)}} \quad \text{and} \quad \frac{\text{collector current (d.c. value)}}{\text{base current (d.c. value)}}$$

are denoted h_{FB} and h_{FE} respectively. In practice at low frequencies h_{fb} is approximately equal to h_{FB} and h_{fe} is approximately equal to h_{FE}. Other '*h* parameters' are described in appendix 3.

A parameter increasingly used with the FET is the transfer conductance g_{fs} (see figure 6.3d) which is defined as

$$g_{fs} = \frac{\text{small change in drain current}}{\text{small change in gate voltage}} \quad \text{(with drain voltage constant)}$$

As can be seen it is similar to the g_m of a valve. Other '*g* parameters' are discussed in appendix 3.

6.3. Load Lines

The performance of a resistance loaded amplifier such as the one illustrated in figure 6.2 can be predicted by superimposing a *load line* on the *output characteristics* of the valve or transistor. The output characteristic is the current-voltage curve plotted for the particular type of amplifier chosen. For the common cathode circuit it is the anode current/anode voltage graph; for the common emitter circuit, the collector-current/collector-voltage graph; and for the common source circuit, the drain current/drain voltage graph. To explain the use of the load line, the valve I_A/V_A graph will be used, as a valve amplifier was described earlier; however, the theory applies equally to either type of transistor amplifier.

As was stated in section 6.1, the equation describing the operation of a simple resistance loaded amplifier such as the one illustrated in figure 6.2 is

$$V_S = I_A R_L + V_A$$

where V_S is the supply voltage, V_A is the anode voltage, I_A is the anode current, and R_L the load resistance. This equation, when rearranged, can be written

$$I_A R_L = V_S - V_A;$$

that is, load voltage = (supply voltage) – (valve anode voltage). Dividing both sides by R_L, the load resistance, gives

$$I_A = \frac{V_S}{R_L} - \frac{V_A}{R_L}$$

For a particular amplifier, supply voltage V_S and load resistance R_L are both constant, so this equation indicates how the anode current varies with anode voltage *when the load is connected.* The I_A/V_A curves show how anode current and voltage vary for the valve on its own. If the graph of I_A/V_A is plotted from the equation given the result is a straight line called the load line, as shown in figure 6.4. The values of I_A, V_A and V_G can now be read off from the graph where the line cuts the valve characteristics. To draw the load line only two points are needed. One is obtained by considering the valve when cut off. At this point I_A is zero, there is no voltage drop across the load and the anode voltage equals the supply voltage. The reason for the valve being cut off is that the grid voltage is very negative as shown in the figure. I_A is zero along the V_A axis so the point where the load line cuts the V_A axis is the cut-off point as shown. It is obtained by putting $V_A = V_S$. The other point is obtained by considering where the load line cuts the I_A axis. At this point the anode voltage is zero and all the supply voltage is dropped across the load resistor to give an anode current of V_S/R_L. Thus, this point is obtained by putting $I_A = V_S/R_L$ on the I_A axis. Joining the points gives the load line. In practice a valve cannot operate beyond point W shown on the curves in figure 6.4 because beyond this point the grid is positive and takes current thus reducing the anode current. Some types of transistor can operate beyond point W; this is considered in the article on bias circuits.

Where the load line cuts the valve characteristics gives the values of anode voltage between which the valve will swing for a given change in grid voltage. For example, in the curves of figure 6.4, when the grid voltage is at –2 V

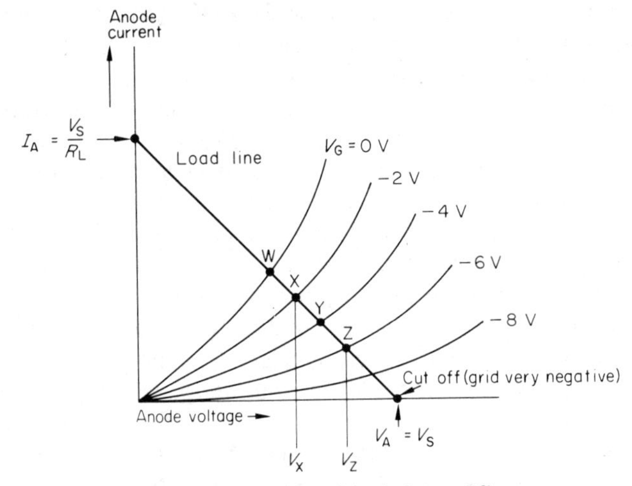

figure 6.4 Load line for triode amplifier

(point X) the anode voltage is V_X as shown. When the grid voltage is at –6 V, the anode voltage is V_Z. Thus, if an input signal makes the grid voltage swing between –2 V and –6 V the anode voltage swings between V_X and V_Z. The anode voltage *change* is thus $V_Z - V_X$ for a 4 V change at the grid. The amplifier gain is thus $(V_Z - V_X)/4$. In practice this value is less than μ (the amplification factor of the valve) because of the a.c. resistance of the valve (r_a). Where R_L is very large compared with r_a, the actual gain can become quite close to the valve amplification factor. The valve can be likened to an a.c. generator (see figure 6.5) generating μV_g volts, where V_g is the grid *a.c.*

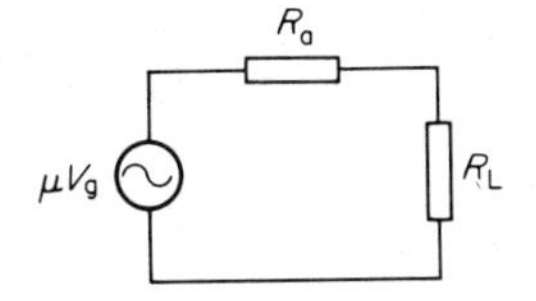

figure 6.5 Triode equivalent circuit

signal, which is fed to two resistors, r_a and R_L, in series. Some of the available voltage μV_g is 'lost' across r_a, the output voltage across R_L being the difference between μV_g and the voltage drop across r_a. Only if r_a is small, and thus the 'lost' voltage small, can the output voltage be comparable to μV_g. An equivalent circuit can be drawn for other active devices and it can be similarly shown that the valve or transistor gain is not actually achieved in practice.

The points where the load line cuts the I_A/V_A curves can be used to give a graph of I_A/V_G for the amplifier with load. This differs from the I_A/V_G curves for the valve alone in that each such I_A/V_G graph is drawn relative to a *constant* value of V_A. In a practical circuit I_A, V_G and V_A are all varying and, whilst at any point the values of I_A, V_G and V_A will be given from the valve curves, any single curve does not give the true picture. If points W, X, Y, Z of figure 6.4 are transferred to a set of I_A/V_G axes as shown in figure 6.6 this gives a single graph, the *dynamic mutual characteristic,* which does tell us how I_A, V_G and V_A are varying all the time. The construction of this curve is important and should be carefully noted. First draw the V_G axis to the left of the existing $I_A V_A$ axes, and scale it in V_G volts. Taking points W and X as illustrations: at point W, $V_G = 0$, I_A has the value shown on the I_A axis; a horizontal line is drawn from W to cut the I_A axis at which $V_G = 0$. This then is point W′ on the dynamic curve. Similarly, at point X, $V_G = -2$ V and I_A has the value shown for $V_G = -2$ V on the I_A axis. Again, a horizontal line is drawn through the I_A axis to meet $V_G = -2$ V. Where the line meets the vertical drawn at $V_G = -2$ V is point X′ on the dynamic curve. The process is repeated for the other points to give the complete curve. The dynamic mutual curve is extremely important since its shape determines how

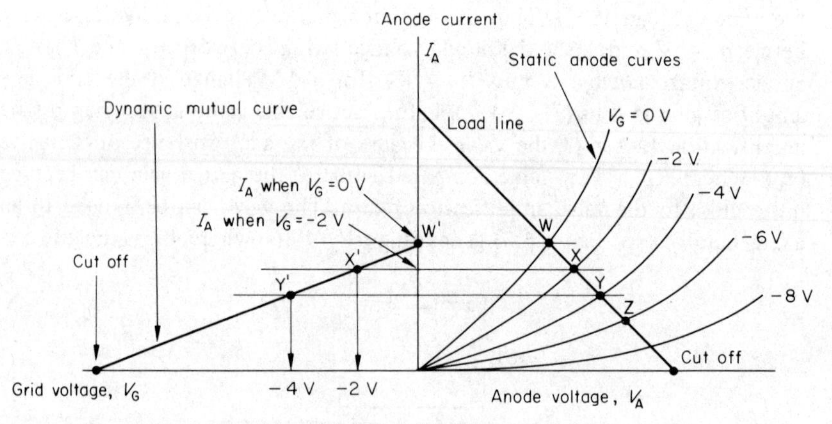

figure 6.6 Construction of dynamic mutual characteristic

well the valve amplifies without distortion. This curve is the transfer curve of the single stage amplifier as a whole and will be the one used to illustrate biasing in the next article.

It is again emphasised that the above theory applies equally to both valves and transistors, the valve merely being used as an example. Thus the dynamic transfer curve of the bipolar transistor could be a plot of collector current/base current (or collector current/base emitter voltage) and for the unipolar transistor could be a plot of drain current/gate voltage, the dynamic curve being obtained in each case by using the load line as described.

6.4. Biasing

Biasing a valve or transistor means selecting a *no-signal* or *quiescent operating point* on the characteristics and setting up a circuit to establish the necessary conditions of voltage and current. The position of the no-signal operating point is determined by the allowable distortion of signal. In the explanation of the simple triode amplifier earlier in the chapter it was stated that as the grid voltage is made more negative the anode current falls and as the grid voltage is made less negative the anode current rises. If the grid is at zero volts when no signal is present (that is, the quiescent grid voltage is zero) a positive signal swing causes the grid to be positive with respect to the cathode and grid current flows. The anode current will not then increase because the grid is taking the electrons that would otherwise have gone to the anode. The result is that for a symmetrical input voltage swing (that is, the positive excursion equals the negative excursion) the anode voltage will not swing symmetrically but will *clip* on the half-cycle caused by the positive grid excursion. This is shown in figure 6.7 which shows the output current change for the same

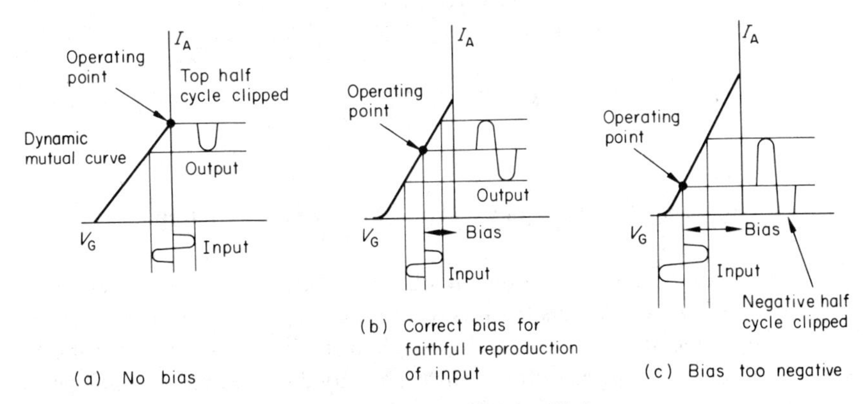

figure 6.7 Effect of bias on fidelity

input signal swinging about various quiescent points. Figure 6.7a shows positive clipping. If the quiescent point is lower down the dynamic transfer curve, as in figure 6.7b (that is, if the grid is held at a *negative bias voltage* when no signal is present), clipping does not occur. If the grid is too negative at the quiescent point, as shown in figure 6.7c, the positive excursion is faithfully copied at the output but the negative excursion causes cut-off, anode current ceases and clipping takes place on the other half-cycle. The *bias voltage* on the grid must be at the right point if distortion of the signal is not to occur.

Similar considerations apply to transistor amplifiers except perhaps FET amplifiers for very small signals. As was explained earlier the bipolar transistor must be forward-biased between base and emitter for correct operation. Again, the wrong value of bias may cause excessive base current on one half-cycle of the input or cut-off on the other half-cycle. FET amplifiers may be operated without bias for very small signals provided that in the case of the JUGFET the gate-channel junction does *not* go into forward bias.

Another point to be taken into account when selecting a bias point is the *linearity* or 'straightness' of the dynamic transfer curve. This curve is not usually a straight line–which is necessary for zero distortion–and the choice of the bias point too near the lower curved region may result in distortion as shown in figure 6.8.

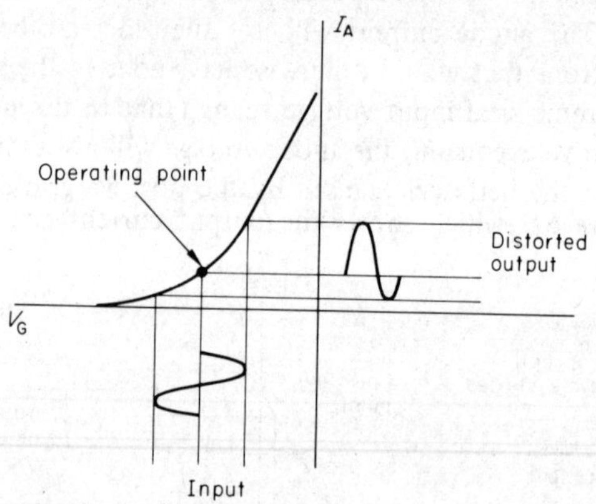

figure 6.8 Effect of transfer curve non-linearity on fidelity

Letter symbols are used to denote *classes of bias,* which are illustrated in figure 6.9. Class-A bias puts the quiescent point approximately in the centre of the dynamic transfer curve and gives distortionless amplification. Class-B biases at cut-off giving an output which is heavily distorted on one half-cycle, and class-C results in distortion of both half-cycles. Class-AB is a combination

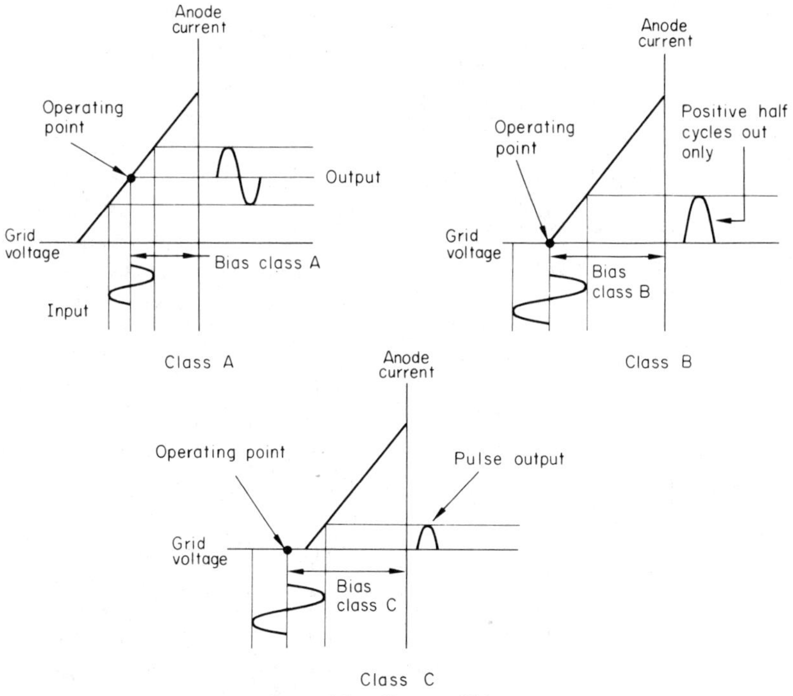

figure 6.9 Classes of bias

of class-A and class-B bias. For an amplifier having minimum distortion Class-A is the obvious choice but class-B may be used in certain circuits where two valves or transistors work 'back to back'. This circuit is described later. Class-C bias is not used in amplifiers but is useful in oscillator circuits as described later.

A selection of commonly used bias circuits is shown in figure 6.10. Figure 6.10a shows cathode bias. In this circuit the valve provides its own bias in that the quiescent anode current is passed through resistor R_K, the voltage drop across R_K shown as V_K being the bias voltage. This is so since the grid is at zero volts with respect to the negative supply line, referred to as earth, the cathode is at $+V_K$ volts with respect to earth. Thus the cathode is at $+V_K$ volts with respect to grid or alternatively the grid is at $-V_K$ volts with respect to cathode. The quiescent anode current is the current flowing with a bias voltage of V_K volts as shown on the dynamic transfer curve diagram given alongside figure 6.10a. This form of bias can only be used where there is a quiescent anode current. Reference to figure 6.9 shows that this only occurs in class-A or class-AB bias. The capacitor C_K in the figure is called a *decoupling capacitor.* Without it, *negative feedback* is introduced and the

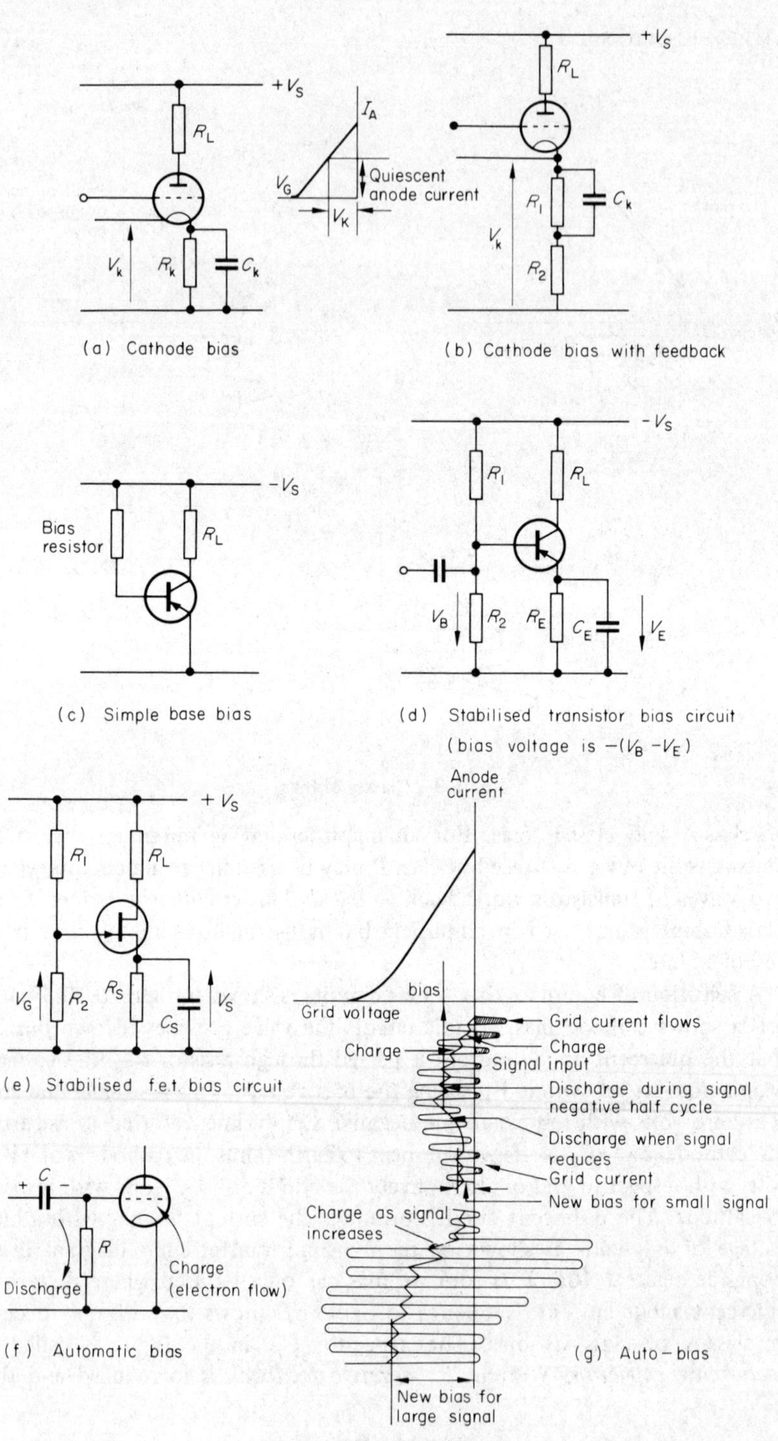

figure 6.10 Bias circuits

amplifier gain is reduced. The reason for this is as follows. When a signal is applied between grid and earth the anode current fluctuates which in turn causes V_K to fluctuate in sympathy with the input signal. V_K can now be considered to consist of a steady direct component and a changing component. The signal which is amplified is that *between grid and cathode* and thus is *the difference* between the input signal and the changing component of V_K. Fluctuations in V_K can be reduced by inserting C_K which, provided it is of the correct value, offers a low reactance path to a.c. The alternating voltage component across C_K is thereby reduced. Insertion of C_K does not, of course, alter the d.c. value of V_K so that the bias voltage remains approximately constant. As the changing component of V_K is more or less removed, the input signal is not reduced and the full value appears between cathode and grid. If feedback is required as a means of improving stability (see chapter 2) *part* of the bias resistor may be decoupled as shown in figure 6.10b. In this case $R1$ and $R2$ form the total bias resistance, $R2$ providing feedback as well as part of the bias.

For the bipolar transistor *bias current* must flow in the control electrode. A simple way of providing a path for such a current is shown in figure 6.10c which shows a simple common emitter amplifier. This circuit is of little practical use if the amplifier is to be stable over a reasonable temperature range. The reason for this is that the collector-base reverse bias sets up a saturation current which is dependent on temperature. The transistor amplifies this reverse current thus increasing the collector current. The increased collector current further increases the transistor temperature which increases the saturation current. The process is *cumulative* and if unchecked the transistor will become permanently damaged. The process is called *thermal runaway.* A commonly used bias circuit to overcome the problem is shown in figure 6.10d. The emitter resistor provides *d.c. feedback,* in that if the collector current rises due to temperature the voltage across it increases, with the result that the bias voltage (which is the difference between the base-earth voltage and the emitter-earth voltage) is reduced, thereby compensating for the increased collector current. To prevent the same feedback in response to fast (a.c.) changes in collector current, the emitter resistor may be decoupled, in a similar manner to the cathode-bias resistor of the valve circuit, by capacitor C_E. The capacitor allows slow (d.c.) changes resulting from temperature changes and so d.c. feedback is permitted. The potential divider $R1$ and $R2$ holds the direct voltage between base and earth fairly constant provided the base current is a small proportion of the current through $R1$, $R2$.

Amplifiers for small signals using field-effect transistors may be operated without bias because of the method of operation of these transistors. For larger signals bias may be provided by a source resistor, the circuit then being similar to figure 6.10a for the valve. If a larger value of source resistor is required to improve d.c. stability the circuit of figure 6.10e may be used. Here, the resultant bias voltage is the difference between V_G and V_S. Part of the increased voltage across R_S is thus 'backed off' by V_G. The increased value of R_S improves stability as in the bipolar circuit. The temperature problem is not, however, as serious with field-effect transistors as the FET does not amplify the saturation current.

Automatic bias is shown in figure 6.10f and 6.10g. Here the signal itself establishes the bias level. On positive half cycles of signal the grid-cathode of the valve behaves as a diode and grid current flows charging the capacitor as shown. On negative half cycles, the capacitor discharges via resistor R but not by the same amount, because the resistance on discharge is high whereas the resistance on charge (the forward resistance of the grid-cathode diode) is low. Thus, C retains some charge and the grid is held negative. The bias voltage at the grid being dependent upon the capacitor, charge builds up over the first few cycles of signal until a point is reached where the charge lost on discharge during the negative half cycle equals the charge gained during the positive half cycle. (The charge gained during the positive half cycle is progressively reduced as bias builds up, because the grid is positive for progressively shorter periods.) Should the signal amplitude fall, the positive excursion does not make the grid go positive and capacitor C discharges until the bias point is reached where the grid can go positive, charging current can flow and equilibrium (that is, charge gained equals charge lost) is restored. Similarly, if the signal amplitude increases, the grid goes positive for a longer period and charge current increases, thus increasing bias and shifting the signal to the left in figure 6.10g again until equilibrium is reached. Auto bias thus depends upon signal strength. It is used mainly in oscillators and in some mixer circuits.

6.5. Coupling in Multistage Amplifiers

The simple resistance loaded triode amplifier considered earlier provided only one *stage* of amplification. In practice many such stages are joined together so that the output from one stage provides the input to the next. This method of forming a *multistage amplifier,* which is called *cascading,* provides an overall gain that is much greater than that of a single stage. The overall gain is, in fact, the product of the individual stages that make up the amplifier.

All valves and transistors require d.c. supplies in order to work correctly. In an a.c. amplifier the a.c. signal is superimposed on a d.c. level; that is, it is produced by a d.c. level that varies. In such an amplifier only the variation (that is, the a.c. signal) is passed on from stage to stage, and the d.c. level at the output of one stage is *blocked* from the input to the next stage. The *coupling circuit* between stages must therefore be capable of allowing the a.c. signal to pass while at the same time not allowing the d.c. level to pass. D.C. amplifiers, on the other hand, which are built to amplify a very slowly varying d.c. level, must have coupling circuits that allow passage of d.c. levels. In general, coupling circuits must offer as little impedance as possible to the particular signal being transferred.

The most common forms of coupling circuit include resistance-capacitance coupling, transformer coupling and direct coupling (see figure 6.11). Resistance-capacitance coupling (shown in figure 6.11a) consists of a resistance loaded amplifier stage capacitively coupled to the next. The coupling capacitor C_C blocks the d.c. level existing at the output of the one stage from the input to the next, but allows the a.c. signal to pass, offering it an opposition that is dependent on frequency. At low frequencies the opposition to signal due to the reactance of the coupling capacitor is high and at high frequencies it is low. A part of the output signal voltage is developed across the coupling capacitor, the remainder being transferred to the input of the next stage. At low frequencies when the coupling capacitor reactance is high, the signal voltage developed across the coupling capacitor is also high

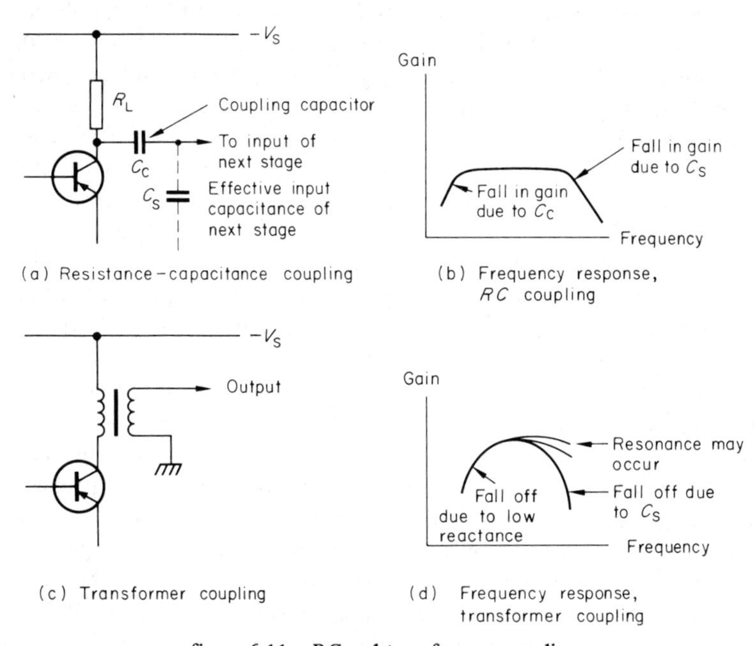

figure 6.11 RC and transformer coupling

and the input signal to the next stage is thereby reduced. Thus, the overall gain is reduced at low frequencies as shown in the gain-frequency curve of figure 6.11b. At the middle and higher regions of the frequency band the effect of the coupling capacitor on gain is very small. However, at the upper end of the range the gain again falls off for a different reason. The input impedance of the next stage should be as high as possible to develop maximum signal voltage across it. The input capacitance between electrodes is directly across the input and so forms part of the input impedance. As the frequency is increased the input impedance is thus reduced (in the same way as the coupling capacitor reactance is reduced) and the input signal falls off. At lower frequencies, where the coupling capacitor reactance is high, the input impedance is high. *RC* coupling is used widely in audio amplifier circuits and may be used, provided that suitable compensation for such fall off in gain is introduced, in amplifiers designed to handle wide bands of frequencies.

Transformer coupling is illustrated in figure 6.11c and the corresponding gain-frequency curve in figure 6.11d. Transformer action passes the variation in output voltage (that is, the signal) to the input of the next stage via the transformer secondary winding. D.C. levels are not passed on and thus one stage is d.c. isolated from the next. At low frequencies the inductive reactance of the transformer, which is the load of the first stage, falls off and gain is reduced. At high frequencies the gain is reduced by input capacitance as explained for *RC* coupling. If the conditions are right, however, resonance between the transformer inductance and input capacitance may occur. This forms a parallel tuned circuit (further discussed in the article on r.f. amplifiers) which has a high impedance. The increased impedance tends to increase gain and may compensate for the falling off which occurs due to the shunting effect of the input capacitance. Transformer coupling is used in a.f. amplifiers and r.f. amplifiers. A.F. transformers have an iron core that contains an air gap to reduce inductance changes due to core saturation; r.f. transformers use a dust or air core since at these frequencies the energy loss in a metal core would be too great.

Amplifiers designed to handle very low frequency variations use direct coupling. With this coupling no frequency sensitive components (capacitors or inductors) are used, because if the frequency is low enough the signal is virtually d.c. and d.c. blocking cannot be permitted. One of the problems of direct coupling is that quiescent d.c. levels are passed from one stage to another and any drift of these levels, due to temperature for example, may be mistaken for a signal and amplified. Special precautions must be taken to avoid such drift amplification in direct coupled amplifiers. Wide-band amplifiers, which are designed to handle signals at frequencies from just above zero to many megahertz, may also employ direct coupling as a means of reducing gain fall off at the lower and upper ends of the band.

6.6. Feedback

Basic ideas of feedback were discussed briefly in chapter 2. As stated in that chapter, the term feedback means taking part (or all) of the output of an amplifier and feeding it back to the input. If the feedback signal assists the input (that is, adds to the input) the feedback is called positive. The feedback signal is then in phase, or has a component in phase, with the input and the apparent gain of the amplifier is increased, because a larger output signal is obtained even though the input applied from *outside* the amplifier remains the same. If the feedback signal reduces the input (that is, the feedback signal is in antiphase, or has a component in antiphase, with the input), the feedback is called negative and the apparent gain falls, as the input from outside the amplifier has been reduced.

Feedback can be further divided into *series* or *parallel* feedback and *voltage* or *current* feedback. The different forms are illustrated in figure 6.12. Series feedback means that the input signal is connected in series with the feedback signal; parallel feedback means that the input signal and feedback

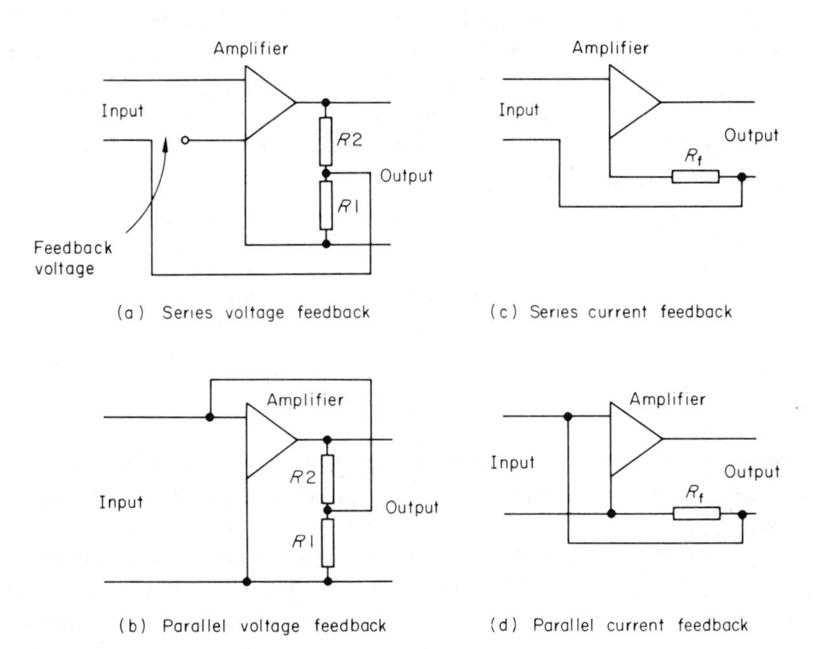

figure 6.12 Forms of feedback

signal are in parallel. Voltage feedback indicates that part (or all) of the output voltage is fed back; current feedback means that the feedback signal is proportional to the output current. In figure 6.12a a potential divider is connected across the amplifier output and a fraction, namely $R1/(R1 + R2)$, of the output voltage is fed back to the input so that it is in series with the input signal. Thus, the feedback is series-voltage, and, as the resistor R1 is in series with the input impedance of the amplifier, so the effective input impedance is increased. The potential divider being across the output, is in parallel with the amplifier output impedance and thus the effective output impedance is reduced. Both these changes of impedance are characteristic of series-voltage feedback.

Parallel voltage feedback is shown in figure 6.12b. Here a potential divider is again connected across the output, reducing the effective output impedance and providing a fraction $R1/(R1 + R2)$ of the output voltage as the feedback signal. On this occasion, however, the feedback signal is in parallel with the input signal and the feedback resistor $R1$ shunts the amplifier input impedance thereby reducing the effective input impedance. Again, this impedance change is characteristic of the type of feedback.

Current feedback is shown in figures 6.12c and 6.12d. The feedback resistor R_f is in series with the output of the amplifier, thus increasing the effective output impedance; and the voltage developed across R_f, which is determined by the output current, is the feedback signal. The feedback signal is in series with the input in figure 6.12c and in parallel with the input in figure 6.12d. As before, series connection adds to the effective input impedance and parallel connection reduces the input impedance. A summary of these effects is as follows.

Feedback	Input impedance	Output impedance
series-voltage	increased	reduced
parallel-voltage	reduced	reduced
series-current	increased	increased
parallel-current	reduced	increased

As can be seen, series feedback always increases the input impedance, parallel feedback reduces it. Voltage feedback reduces the output impedance, current feedback increases it.

Practical circuit examples are shown in figure 6.13. Series-voltage negative feedback is shown in figure 6.13a. Here, the potential divider made up of registers $R1$ and $R2$ is connected across the output, and the feedback signal, which is $R1/(R1 + R2)$ of the output is developed across $R1$. The feedback is in series with the input, because the total signal between grid and cathode is made up of the input signal and the p.d. across $R1$. The signal p.d. across the bias resistor R_K can be ignored, for R_K is bypassed. The feedback is negative

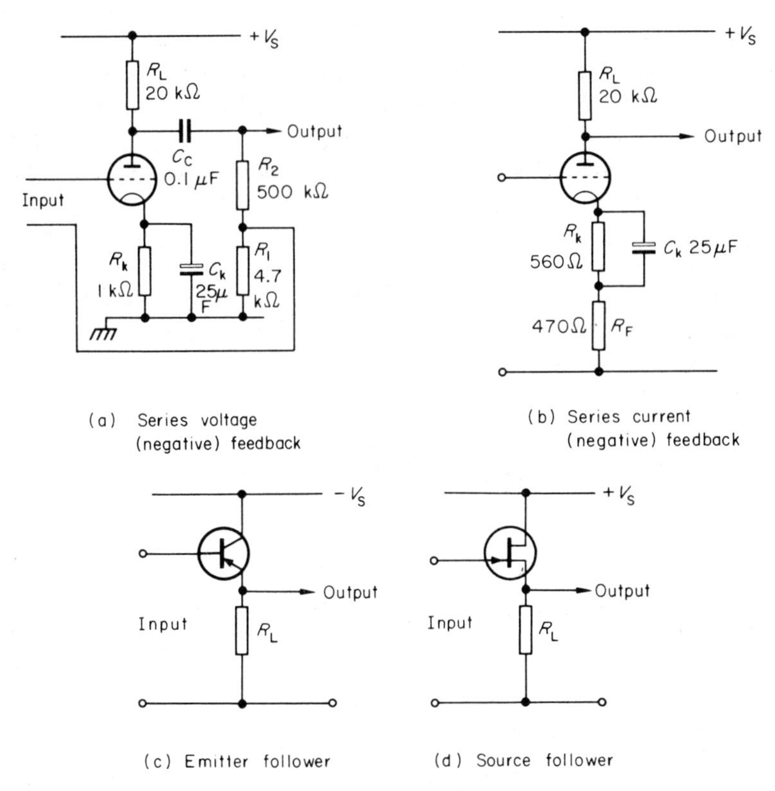

figure 6.13 Feedback circuits

because of the phase-shift of 180° (that is, as the input signal rises, the output signal falls) of the resistance loaded amplifier; this phase shift puts the feedback signal in antiphase with the input signal, thereby reducing it.

A circuit using series-current negative feedback is shown in figure 6.13b. In this circuit part of the bias resistor through which the valve current is passing, is not bypassed and an alternating (signal) voltage is developed across it. The actual input signal amplified is the *difference* between the grid-earth input signal and the feedback signal developed across R_F. Two circuits using 100% series-voltage negative feedback are shown in figure 6.13c and d. In these circuits, the load is connected between emitter (or source) and earth, and the entire output voltage is fed back so that the actual input to the transistor is the difference between the base-earth (or gate-earth) signal and the output voltage. As was stated earlier, these circuits are the common-collector and common-drain circuits, otherwise known as the emitter-follower and source-follower respectively. The valve equivalent circuit (not shown) is the cathode-follower. All 'follower' circuits have high input impedance and low output impedance due to the high percentage of series voltage feedback and, as the output voltage rises when the input voltage rises, the phase shift is zero. There is no voltage gain as the output voltage is less than the input voltage; the circuits are widely used however for *impedance matching.* Impedance matching was discussed in chapter 4, the device then mentioned being the transformer. Follower circuits are most useful when it is required to match a high output impedance to a low input impedance. By inserting a follower circuit, effective matching is achieved, as the high output impedance matches the high follower input impedance and the low follower output impedance matches the low input impedance of the succeeding stage. The technique is shown in figure 6.14.

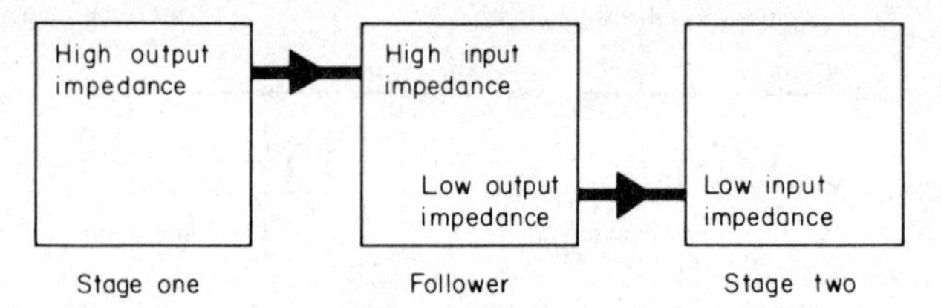

figure 6.14 Use of follower circuit in impedance matching

6.7. Small Signal Audio Amplifiers

Figure 6.15 shows typical circuits for a two stage *RC* coupled audio amplifier; part (a) shows a valve circuit using pentodes and part (b) shows a transistor circuit using bipolar transistors.

In figure 6.15(a) resistors *R*2 and *R*4 are the load resistors for the first and second stage of amplification respectively; *R*3 and *R*7 are the cathode bias resistors for class-A operation, and these are bypassed (decoupled) by capacitors *C*3 and *C* 7 respectively. Capacitors *C*1, *C*2 and *C*5 are coupling capacitors, which allow a.c. signals to pass but block d.c. levels of one stage from the next. *R*6 is the *grid-leak* resistor that provides a discharge path for the grid of valve *V*2 and also develops (across it) the input signal to *V*2.

Screen grid current of *V*1 and *V*2 flows through resistors *R*1 and *R*5 respectively and the screen voltage is thus the HT supply voltage less the drop across these *screen resistors.* To prevent variation of the valve screen voltages due to changing valve current, screen grids of *V*1 and *V*2 are decoupled by capacitors *C*6 and *C*4 respectively, the action being similar to cathode decoupling capacitors discussed earlier.

Figure 6.15b shows a two stage *RC* coupled bipolar transistor amplifier. Amplification takes place in transistors *Tr*1 and *Tr*2, the respective loads

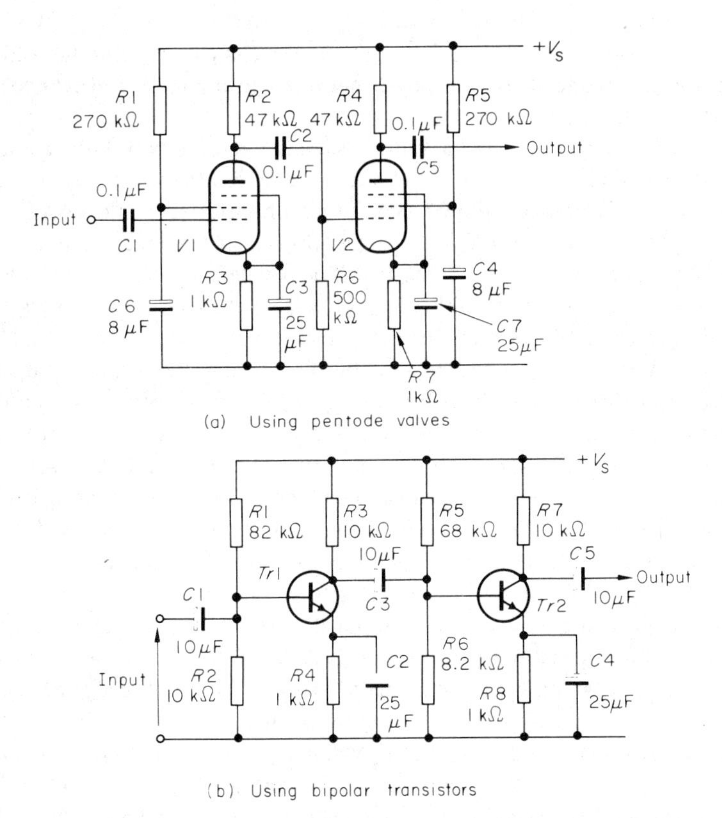

figure 6.15 Two-stage audio amplifier

being resistors $R3$ and $R7$. Resistors $R1$, $R2$ and $R4$ form the bias circuit of $Tr1$ and resistors $R5$, $R6$ and $R8$ the bias circuit of $Tr2$. Capacitors $C1$, $C3$ and $C5$ are coupling capacitors and emitter bypass capacitors are $C2$ and $C4$ for $Tr1$ and $Tr2$ respectively.

6.8. Audio Power Amplifiers

The main requirement of an audio power amplifier is that maximum power at audio frequencies should be transferred from any previous small-signal voltage amplifier to the loudspeaker driven by the power amplifier. Power amplifiers are thus large-signal amplifiers. The important features of such amplifiers are *dissipation, efficiency, distortion* and *matching.*

Because of the high power level the valve or transistor must be able to handle larger values of voltage and current than is necessary for small-signals.

and it is important to make sure that the power dissipated *within* the valve or transistor (as distinct from the power delivered to the loudspeaker) does not exceed the safe value.

The efficiency of a power amplifier is a measure of how well the amplifier converts the d.c. power supplied to the valve or transistor into useful audio output power. It can be shown that for an amplifier using class-A bias, the efficiency can never exceed 50%. If class-B bias is employed, the maximum available efficiency rises to 78.5%. It will be recalled that class-B bias sets the operating point at cut-off (figure 6.9) and if this class of bias is used something must be done to reduce the resultant distortion.

A non-linear transfer curve (relating output signal waveform to input signal waveform) produces a distorted waveform containing harmonics. For small signals a small linear part of the overall curve may be used; in large signal amplifiers, however, the problem is more serious, because a larger part of the transfer curve is in use and the curved portions at top and bottom may affect the output waveshape. The smallest amount of distortion normally detected by the ear is when the harmonic content is 5% of the fundamental.

It is essential for a power amplifier to deliver maximum power to the output. This occurs only when the input impedance of the circuit fed by the power amplifier, usually a loudspeaker, is equal to the output impedance of the amplifier. Matching can be achieved using an output transformer of the appropriate turns ratio (see chapter 4: transformers).

A class-A power output stage is shown in figure 6.16. The circuit employs a valve but a transistor could equally well be used. Bias is provided by resistor $R1$ decoupled by capacitor $C1$. The anode load consists of an output transformer that matches the stage to the loudspeaker. The circuit is called a single-ended power output stage because only one valve is used. As was explained earlier, class-A bias requires a quiescent (no-signal) standing

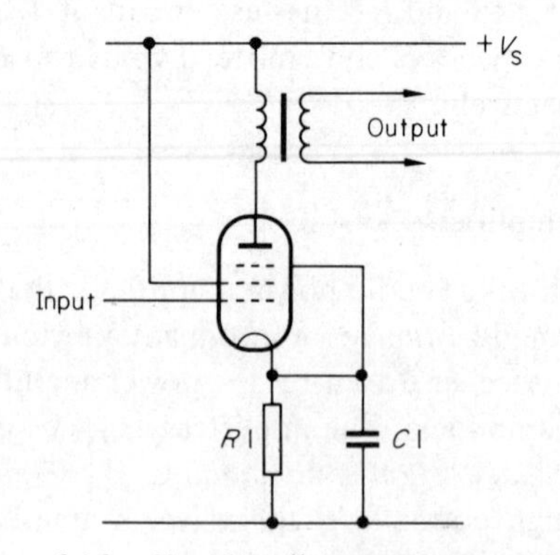

figure 6.16 Class A audio power output stage

current. In small signal amplifiers the quiescent levels are small and power dissipation low. In power amplifiers, however, the quiescent power dissipation is high and the current drain in battery-supplied circuits may be a problem. To avoid this, class-B biased circuits may be employed.

A push-pull output stage is illustrated in figure 6.17a. This circuit employs bipolar transistors but valves may also be used. The circuit consists of two similar transistors supplied with equal but antiphase input signals as shown at inputs 1 and 2. The transfer curves of each transistor, together with the

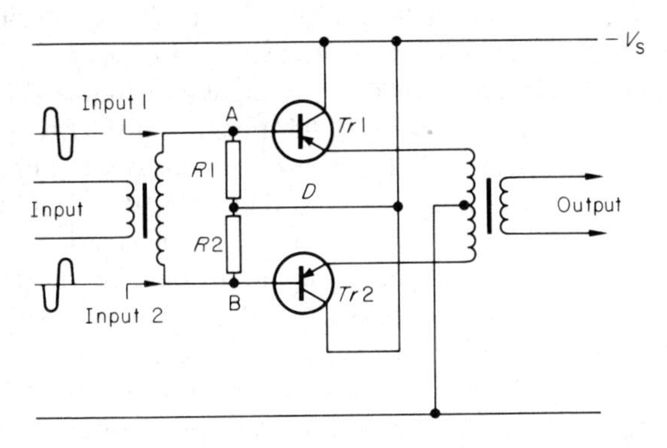

(a) Push-pull output stage using transformer drive

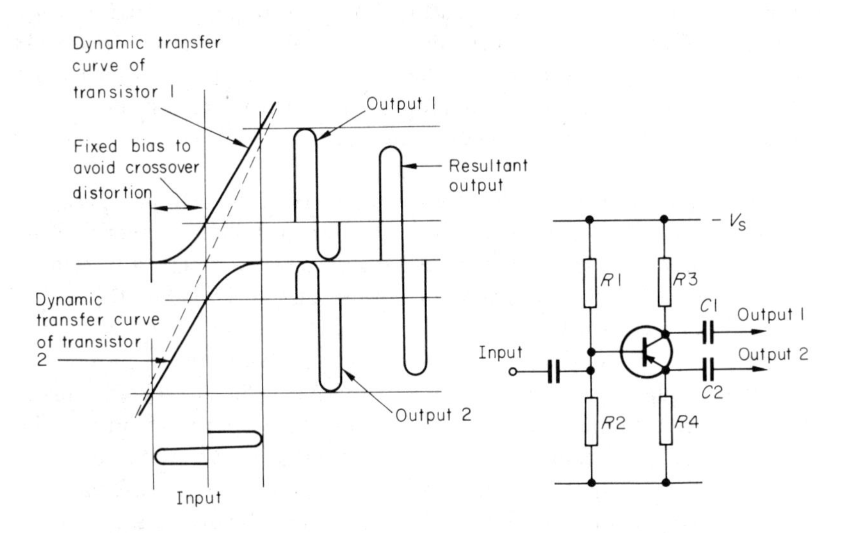

(b) Overall transfer curve for push pull output stage, class-B

(c) Phase splitter

figure 6.17 Push-pull circuits

composite or overall transfer curve, are shown in figure 6.17b. Each transistor amplifies its own input signal (as shown) and, if class-B bias is used, alternate halves of the input are amplified by each transistor to give outputs 1 and 2 as shown in the figure. The two outputs are summed in the output transformer to give the single resultant output shown. Because of the non-linearity of the individual transfer curves distortion would occur if the transistors were biased actually at cut-off. Such distortion is called *cross-over* distortion. To avoid cross-over distortion, a small forward bias is applied so that, strictly speaking, the bias is class-AB. Nevertheless, high efficiencies are available with such circuits. Bias is provided in the circuit shown by resistors $R1$ and $R2$. The two equal antiphase inputs are provided by an input transformer connected across $R1$ and $R2$ in series. The midpoint of this potential divider is earthed as far as signal is concerned (the supply has a low a.c. impedance) and so the voltage variation at input 1 is in antiphase with that at input 2. Consider the instant when the voltage across the input transformer secondary is increasing. The potential at point A with respect to point D would be increasing, as would the potential at point B with respect to point D; but because D is effectively earthed, if the potential at A is positive with respect to D, the potential at B would be negative (that is, below earth potential) with respect to D. Thus the voltage changes at A and B with respect to D are in antiphase. An alternative push-pull driver circuit, called a *phase-splitter* is shown in figure 6.17c. This consists of a normal amplifier circuit without a decoupling capacitor. Thus, there is an alternating component developed across the emitter resistor. As the collector current rises, the emitter voltage rises as the collector voltage falls. Thus outputs 1 and 2 are in antiphase. Adjustment of resistors $R3$ and $R4$ ensures equal output signals. Capacitors $C1$ and $C2$ are coupling capacitors.

Apart from the obvious advantage of using class-B or -AB bias, push-pull circuits have a second important advantage over single-ended stages that use output transformers. The standing current associated with class-A bias passes through the output transformer primary and sets up a standing flux in the core. Further increases in current due to signal may force the transformer into *saturation,* after which flux can no longer increase. As the flux cannot change, the induced output voltage cannot change, and the voltage remains constant once saturation is reached. This clips the output waveform. Audio transformers usually contain an airgap to prevent such saturation occurring, but the process cannot be taken too far without seriously affecting the correct working of the transformer by reducing the effectiveness of the iron core. Examination of the push-pull circuit, however (figure 6.17a) shows that the standing current of transistor $Tr1$ flows in the opposite direction in the transformer primary to that of the standing current in transistor $Tr2$. Thus, the standing fluxes cancel and saturation, with its consequent distortion, is therefore avoided. This 'opposition' does not, of course, prevent changing fluxes having an effect, because the changes due to transistors *Tr*1 and *Tr*2 occur on separate half cycles. This cancellation of harmonic distortion helps in another way, for if the power supply is mains derived and contains a hum component at a multiple of the mains frequency, the components cancel in the transformer primary and the hum is not transferred to the loudspeaker.

Figure 6.18a shows a push-pull output stage that does not employ an output transformer. Transistor *Tr*1 is biased by resistors *R*1, *R*2 and *R*3; transistor *Tr*2 is biased by resistors *R*4, *R*5 and *R*6, the a.c. load on both *Tr*1 and *Tr*2 is the loudspeaker coupled by capacitor *C*1. The antiphase inputs are derived from a two-secondary transformer as shown.

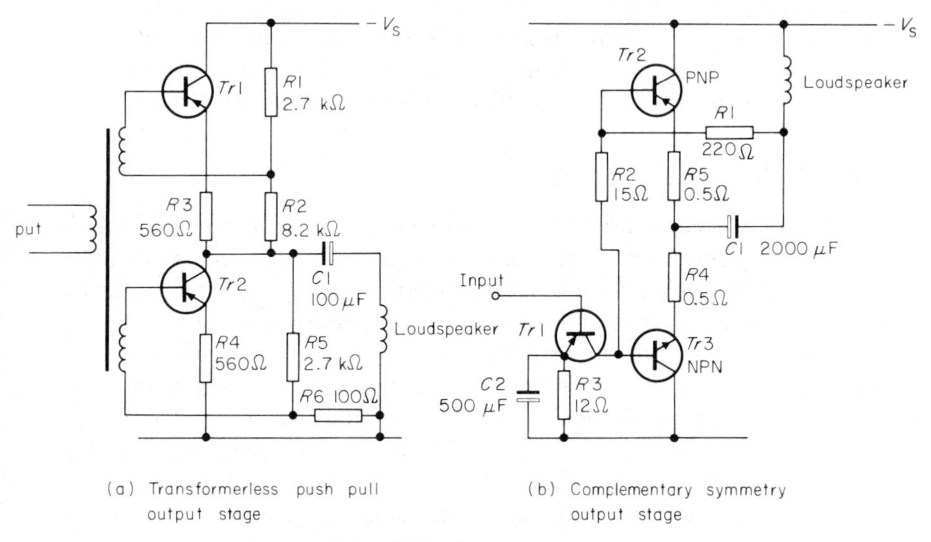

figure 6.18 Output stages

An increasingly popular circuit that uses no transformers is shown in figure 6.18b. It will be recalled that there are two types of bipolar transistor, PNP and NPN, the one requiring voltages of opposite polarity to the other. The *same* input signal applied to a PNP amplifier and to an NPN amplifier, both biased at class-B, would produce individual outputs similar to those shown in figure 6.17b, the outputs being equal if the transistors had similar transfer properties. In the circuit shown *Tr*1 is a driver transistor, biased by *R*3, with a suitable voltage applied to the base, and having a resistive load made up of *R*1 and *R*2. Outputs are taken from the collector of *Tr*1 and the junction, of *R*2 + *R*1. Both outputs are approximately equal and in phase. One output is taken to PNP transistor *Tr*2, which amplifies one half of the input signal; the other output is the input to NPN transistor *Tr*3 which amplifies the other half of the input, thus retaining the natural advantages of push-pull operation. Without *R*2 the circuit would still work but cross-over distortion would occur; *R*2 compensates for the slightly different bias conditions required and non-linearity of the individual transfer curves of the transistors. The resistive load of *Tr*1 acts as the bias path for the two output transistors; resistors *R*4 and *R*5, which are usually of low value, being included to improve stabilisation of the operating point (see section 6.4). The transistors *Tr*2 and *Tr*3 are a matched pair-having similar gain and transfer characteristics. The circuit as a whole is said to employ *complementary symmetry.* If desired, feedback may be introduced via a resistor connected between the junction of *R*4 and *R*5 and the input of *Tr*1. Notice that no coupling capacitors are used in this circuit, other than the one between the output and the loudspeaker; this is made possible by the low working voltages of transistors. The complementary symmetry form of circuit is not possible with valves.

6.9. Radio Frequency Amplifiers

A radio frequency amplifier is a narrow band amplifier, because it is designed to handle a relatively narrow band of frequencies within the r.f. spectrum. The gain/frequency curve of such an amplifier is shown in figure 6.19a. The gain of any amplifier depends to some degree upon the impedance of the load on the amplifier, and to obtain a gain/frequency variation similar to that required, a load having a frequency sensitive impedance is necessary. The impedance of a *parallel tuned* circuit (an inductor and capacitor in parallel) varies with frequency in a manner similar to the required gain/frequency curve, and consequently r.f. amplifiers invariably employ a tuned circuit load as shown in figures 6.19b and 6.19c. Interstage coupling is usually achieved by using the tuned circuit coil as the primary of a coupling transformer. At radio frequencies an iron core has an excessive power loss and dust or air cores are normally used.

The bandwidth of a voltage amplifier is normally defined as the difference between the frequencies at which the voltage gain falls to 0.707 of the maximum gain. If two similar amplifiers are connected in cascade the overall maximum gain, neglecting coupling losses, will be the square of the individual maximum gain. The overall gain at the frequencies where the single amplifier gain is 0.707 of the maximum will be 0.707×0.707 of the overall maximum, that is, 0.5 of the overall maximum. Thus, the overall bandwidth, which is the separation between the frequencies at which the overall gain falls to 0.707 of the overall maximum will be *less than* the bandwidth of a single stage. With

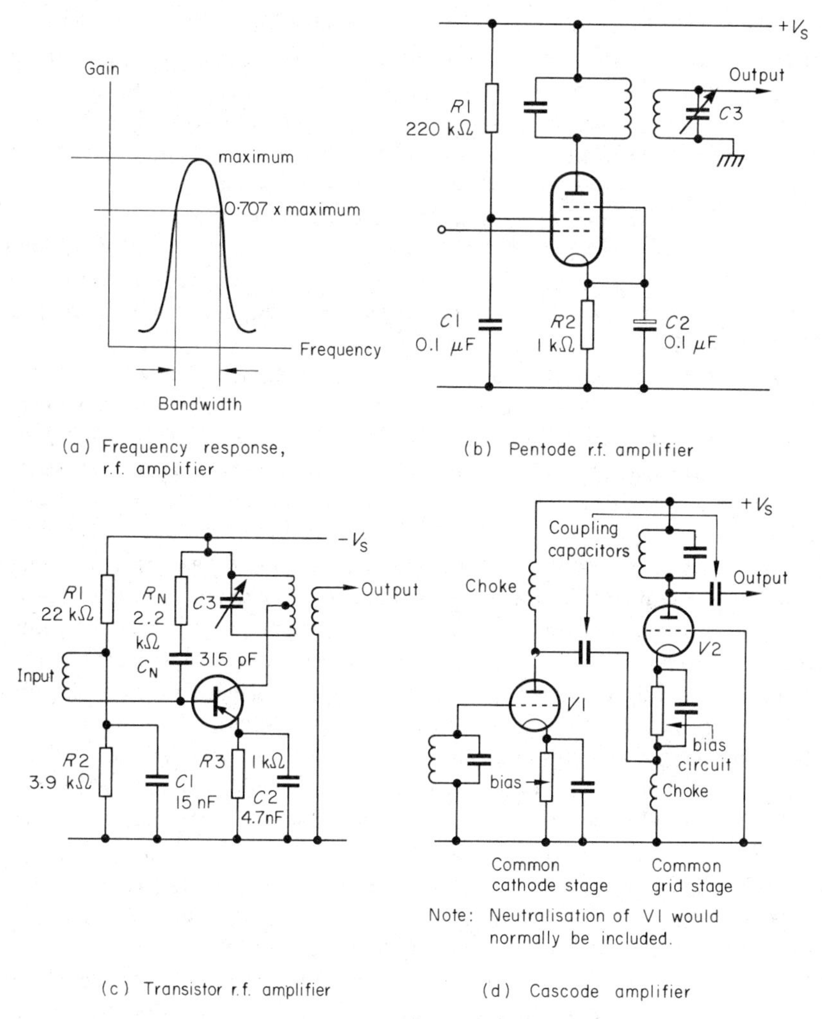

figure 6.19 Radio-frequency amplifiers

several stages the overall bandwidth will be severely reduced, possibly to a point where it is insufficiently large to contain the component frequencies of a modulated signal. To avoid this happening it is necessary to closely control the *coupling* between the individual r.f. stages. The bandwidth may be increased and the gain/frequency curve 'flattened' by one or more of the following methods.

(1) *Band-pass coupling* uses two tuned circuits, for example, by shunting the secondary of the coupling transformer in figure 6.19b by a capacitor, and

loosely coupling the coils. The degree of coupling, that is 'loose' or 'tight', is determined by such factors as the closeness of the coils and their core material, which both affect the amount of magnetic flux linking the coils.

(2) *Staggered tuning* uses two tuned circuits as before but with each tuned to a slightly different frequency. The individual gain/frequency curves combine to give an overall flatter curve centred about a mid-frequency that is between those of the two individual frequencies.

(3) *Damped tuning* uses two tuned circuits that are 'damped' by additional parallel resistors. The impedance at resonance is reduced by such a connection and the gain/frequency curve flattened slightly, thereby increasing the bandwidth.

The circuit of figure 6.19b shows a pentode r.f. amplifier using class-A bias. Resistor $R1$ is the screen grid resistor to set the required screen voltage and $R2$ is the class-A bias resistor. Capacitors $C1$ and $C2$ are bypass (decoupling) capacitors. The circuit of figure 6.19c shows a bipolar transistor r.f. amplifier. Resistors $R1$, $R2$ and $R3$ are bias resistors. Capacitors $C1$ and $C2$ are decoupling capacitors, $C1$ being necessary to ensure that $R2$ is not in circuit as far as the input signal is concerned. The input signal is transformer coupled to the base. The transistor collector is not connected to the bottom of the tuned circuit coil as the relatively low transistor resistance might damp the tuned circuit excessively. (The transistor and tuned circuit are in parallel as far as signal is concerned if the power supply resistance is negligible.) By using the connection shown, only part of the coil is shunted by the transistor. This connection also helps interstage matching between impedances.

$C3$ is the capacitor forming part of the tuned circuit in both figures 6.19b and 6.19c. Altering the value of this capacitor changes the resonant frequency and therefore the midband frequency at which maximum gain occurs. For safety reasons, in the valve circuit (6.19b), $C3$ is not connected in the anode-HT lead but is connected on the secondary side of coupling transformer. The valve load is still a tuned circuit, the impedance being 'reflected' into the anode circuit via the transformer (see appendix 2). In the transistor circuit the variable capacitor can be connected directly in the collector lead because supply voltages are low.

Both circuits show *tunable* r.f. amplifiers as used, for example, in the first stage of a radio receiver. Intermediate frequency (i.f.) amplifiers are of similar form but are not continuously tuned, because the centre frequency (the i.f.) is preset and all circuits are tuned to it. Any variable capacitors present are of the 'trimmer' variety (that is, having a small range of variation) and are used to overcome slight variations in stages that should otherwise be similar. The trimmer capacitors are adjusted until the same performance is obtained from each stage of i.f. amplification.

One problem associated with radio frequency amplification is that of internal feedback through the inter-electrode capacitance between output and input. As was explained in chapter 4, the pentode was developed to reduce the effect of the anode-grid capacitance which causes feedback in a common-cathode amplifier. This is, of course, the reason for the use of the pentode in r.f. amplifiers. With transistors, however, some other way is required to control the level of this undesirable feedback. One circuit to carry out this process, called *neutralisation,* is also shown in figure 6.19c. Feedback is deliberately introduced from the top of the collector coil to the base via R_N and C_N. The phase of the voltages at the top and bottom of the coil is such that the deliberate feedback from the coil top can be used to cancel the internal feedback from the coil bottom.

The feedback due to internal capacitance is severe only at high frequencies in the common cathode, common emitter or common source modes. The common grid, common base or common gate modes do not present such a problem. One method of controlling internal feedback is thus to feed the output from, say, a common cathode stage to a common grid stage (or common emitter to common base, or common source to common gate). Such a circuit is called a *cascode* circuit. A valve cascode circuit is shown in figure 6.19d. The transistor cascode circuit, one form of which is available in integrated circuit form, is similar in layout with appropriate changes for bias circuits etc. (The term 'cascode' should not be confused with 'cascade' which though it includes 'cascode', has much wider application.)

6.10. Low Frequency and Wideband Amplifiers

Wideband amplifiers are those designed to handle a wide range of frequencies ranging from a few hertz to many millions of hertz. Since amplifiers handling square wave signals (for example, video amplifiers in television systems) must be capable of handling the many component frequencies making up a square wave, these amplifiers are considered wideband. Low frequency amplifiers are those designed to handle slowly changing signals at frequencies not much above zero. The problems of low frequency amplification and the methods used to combat them are common to both kinds of amplifier.

The gain/frequency curve of an amplifier using *RC* or transformer coupling between stages falls off at low frequencies, either as a result of the increasing coupling capacitor reactance, or in response to the reducing transformer impedance. In both types of circuit the fall-off is caused by frequency sensitive components.

Two methods are available for counteracting the fall-off in gain. These are, firstly, using more frequency sensitive components to counteract the effects of those causing the fall off; and, secondly, removing the frequency sensitive components and employing *direct coupling* instead.

Circuits employing the first method are shown in figure 6.20. Part (a) shows a wideband amplifier stage using a peaking coil as part of the anode load. At high frequencies the increased impedance of the load increases the

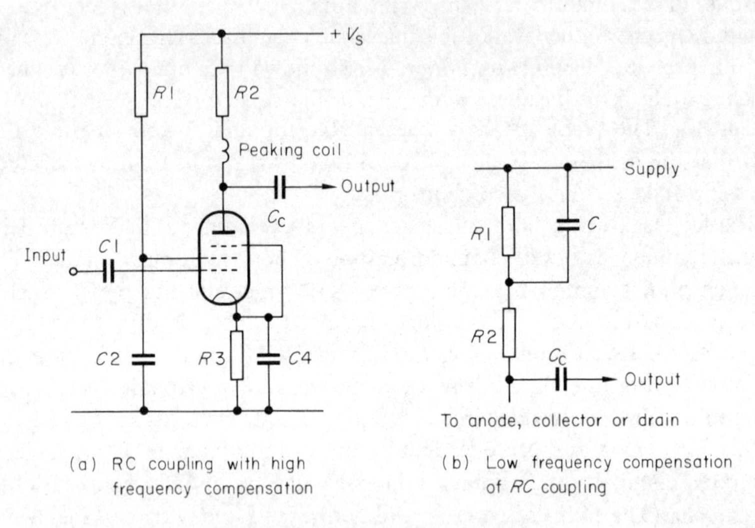

figure 6.20 Low-frequency amplifiers

gain and counteracts the falling-off effect. Occasionally a series inductor is added to the coupling capacitor to form a series-acceptor circuit at the high frequency end of the band, and this too increases the signal passed to the next stage. Figure 6.20b shows one method of improving the low frequency performance of a wideband amplifier. In this circuit, part of the stage load is shunted by a capacitor. At low frequencies the reactance is high and has little effect on the a.c. impedance of $R1$ and C combined. The total load is then approximately $R1$ in series with $R2$. As the signal frequency increases the capacitor reactance is reduced and $R1$ is shunted. The overall impedance falls to give a lower value of stage load over the mid-band frequencies. The gain is higher at low frequencies which tends to compensate for the reduction in signal transfer caused by the coupling capacitor. Both methods may be employed simultaneously in a single video stage.

A circuit using direct coupling is shown in figure 6.21a. If valves are employed care must be taken to avoid excessive grid-cathode voltages; and, because anodes are directly connected to grids, it is necessary to have high cathode-earth voltages (as shown). The values given are typical of those found

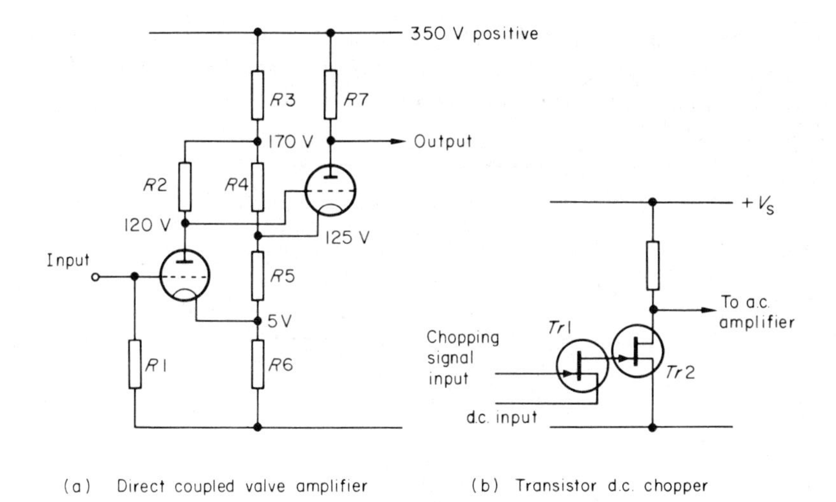

figure 6.21 Direct-coupled low frequency amplifiers

in direct coupled valve stages. Note that the required voltage levels are provided by a potential divider arrangement ($R3$, $R4$, $R5$ and $R6$) connected across the HT supply.

Transistors operate at much lower voltage levels and direct coupling does not involve the provision of such high voltages to reduce bias. However, a further problem exists as solid state devices are prone to d.c. drift as a result of temperature change. In low frequency amplifiers the signals are at frequencies only a few cycles above zero and drift in bias levels may be mistaken for a signal. In an a.c. amplifier two kinds of feedback are employed, one to stabilise d.c. levels, and the other to stabilise a.c. signal gain. The amounts of feedback in each case are usually different. In a direct-coupled low-frequency amplifier feedback to control drift will also control the signal, as it is difficult to distinguish one from the other.

One method of avoiding the problem is to change the d.c. signal into a pulsating signal using a chopper circuit such as that shown in figure 6.21b. Here the d.c. signal is passed through a transistor *Tr*1 that is being switched off and on by a separate chopping signal. The d.c. signal is passed in pulse form to *Tr*2, which sees the signal as an a.c. signal. From *Tr*2 the output can be passed to a compensated a.c. amplifier and suitably amplified. The a.c. amplifier can, of course, distinguish its own a.c. drift from the pulsating signal from the chopper circuit and separate feedback techniques can be employed in the usual fashion. There are various other circuits employed for chopping but the principle is the same for all of them.

Another circuit commonly used to combat the drift problem in both low frequency and wideband amplifiers is the *difference amplifier* shown in figure 6.22. This circuit is the basis of many integrated circuit *operational amplifiers* described later.

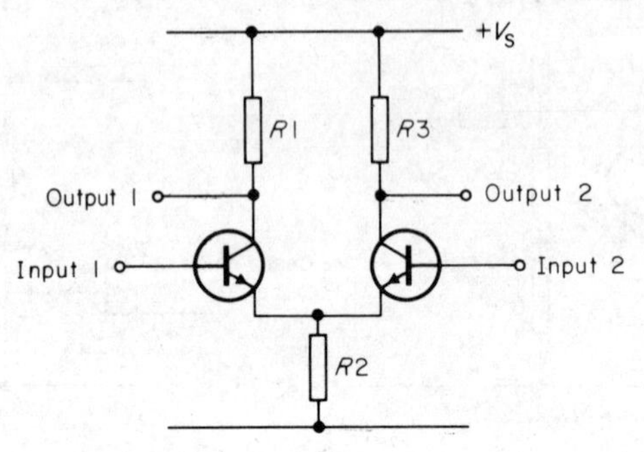

figure 6.22 Simple difference amplifier

The difference amplifier, which is shown using bipolar transistors but may equally employ unipolar transistors or valves, consists basically of two stages with emitters linked as shown. The output is taken from between collectors and there are two input points if required. The output is the *difference* between the individual outputs from the two transistors. If two similar in-phase signals are applied to the inputs the output is zero, because, provided the transistors are *matched* (equal gain etc.), both individual outputs change by the same amount. Operation under these conditions is called *common-mode* operation. If two similar but antiphase signals are applied to the inputs, the output is twice that of one of the individual outputs, because each output changes by the same amount but in the opposite direction to the other. This type of operation is called *differential-mode* operation. The *common mode rejection factor* (or ratio) of a difference amplifier is the ratio between the outputs for differential-mode and common-mode operation for equal input signals. If the valves or transistors are perfectly matched the common mode rejection factor is infinitely high, because common mode operation for equal inputs gives zero output. In practice it should be as high as possible for best operation. The main advantage of this amplifier lies in the fact that drift in input levels as a result of temperature or other causes will be the same for each transistor, and thus as far as drift is concerned the amplifier is operating common mode and no output will result from drift.

If only one input is required the other may be earthed so that one individual output is constant for a changing input signal whilst the other output changes. The overall output, which is the difference between individual outputs as before, is then dependent on input but with the distinct advantage of built-in drift compensation. The type of difference amplifier shown is also known as a 'long-tailed pair'.

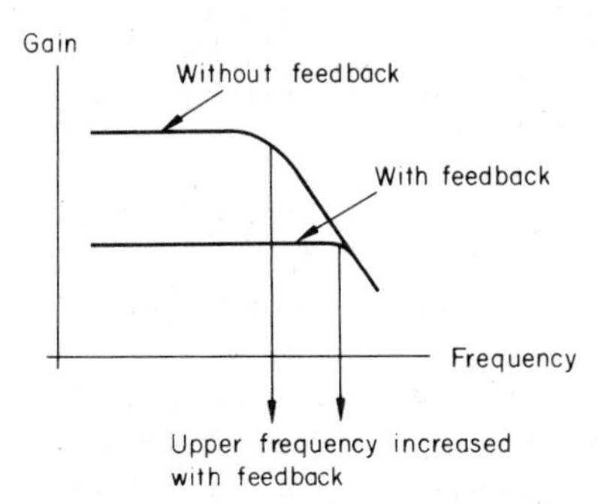

figure 6.23 Increase of upper frequency limit of operational amplifier

Operational amplifiers are very high-gain wideband amplifiers that with certain external circuitry and a feedback loop may be used to carry out mathematical processes such as addition, subtraction, multiplication, division, root extraction, integration, differentiation and so on. Modern integrated-circuit versions use difference amplifier techniques to reduce drift problems and so lower the lower frequency limit of use. The upper frequency limit is considerably increased by the use of negative feedback (as shown in figure 6.23) and often by additional frequency sensitive components connected externally to counteract the effects at high frequencies of the inherent stray capacitances.

7 Oscillation

Oscillators form a basic part of a large number of electronic systems ranging from radio and television transmitting and receiving systems to industrial control and measuring systems. The basic principle of most oscillators, that of positive feedback being applied to an amplifier so that it provides its own input, was considered in earlier chapters. Such an oscillator consists of three parts (shown in figure 7.1a): the amplifier; a network providing feedback and selecting the fraction of output signal fed to the input; and, finally, a frequency sensitive circuit that controls the oscillator output frequency. Many feedback oscillators use the frequency selective properties of a tuned circuit; others use resistor-capacitor phase shifting networks.

Two other basic forms of oscillator are the negative resistance type, in which a component having a rising-voltage falling-current characteristic (that is, a 'negative resistance') opposes the positive resistance of a naturally resonant circuit, and the relaxation type which employs the switching properties of valves or transistors.

Oscillators that use tuned circuits have a sine wave output. Such circuits include feedback oscillators and negative resistance oscillators. Relaxation oscillator output waveforms are non-sinusoidal and may have a sawtooth shape or be of rectangular pulse form.

7.1. Feedback *LC* Oscillators

An inductor and capacitor connected in parallel form a naturally resonant circuit. Energy is passed continually from the capacitor electric field to the inductor magnetic field and back again, the voltage across the parallel combination varying sinusoidally as the transfer takes place. These natural oscillations eventually die away because energy is lost in overcoming the circuit resistance. In a feedback oscillator controlled by a tuned circuit a fraction of the amplified output is fed back to the tuned circuit and replaces

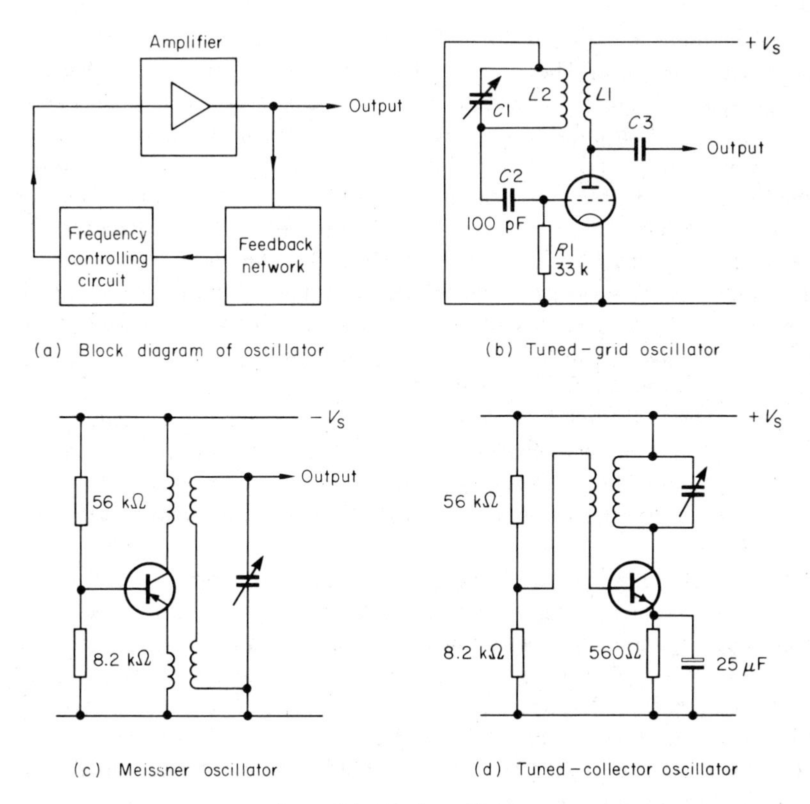

(a) Block diagram of oscillator

(b) Tuned-grid oscillator

(c) Meissner oscillator

(d) Tuned-collector oscillator

figure 7.1 Basic oscillators

the energy lost in the resistance. Thus, oscillations are maintained at the natural frequency of the tuned circuit. This is approximately equal to $1/2\pi\sqrt{(LC)}$ where L and C are the values of inductance and capacitance respectively.

The *LC* feedback oscillators to be examined include Meissner, Hartley and Colpitts oscillators. All feedback oscillators require an amplifier, a feedback signal and a 270° to 360° phase shift round the feedback loop. A feedback signal that is phase shifted by 270° to 360° from the input, may be resolved into two components, one of which is phase shifted by 360°. (As one cycle occupies 360° a phase shift of 360° puts the feedback component in phase with the input.) This component is the one that maintains oscillation.

The circuits given include a variety of transistor and valve circuits, in most of which it is equally possible to use either transistors (including FETS) or valves, provided that appropriate bias arrangements are made.

Figure 7.1b shows a tuned-grid oscillator in which feedback is obtained by mutual coupling between the tuned circuit coil $L2$ and a coil $L1$ that is in series with the valve. The resonant frequency is adjusted by altering $C1$. Components $C2$ and $R1$ determine the bias which is of the automatic type described in section 6.4. Class-C bias is normally used in oscillators. With this class of bias, pulses of current flow in the valve and energy is thus fed into the tuned circuit to maintain oscillation. Distortion of the output does not occur (as it would if class-C bias were employed in, say, an audio amplifier) because the waveform purity is maintained by the natural cycling of the tuned circuit. The circuit shown is a series connected circuit, for the coil transferring energy is in series with the valve. A shunt circuit is also possible in which the transfer coil $L1$ is in parallel with the valve, the valve load in this case being a choke at the frequency generated. The frequency of oscillation is $1/2\pi\sqrt{(LC)}$ where L is $L2$ plus the 'reflected' inductance transferred by the mutual coupling. The loop phase shift is a combination of that caused by the valve (between 90° and 180°) and that resulting from the transformer coupling (180° if connected correctly) making a total phase shift that lies between 270° and 360°.

Figure 7.1c shows a true Meissner circuit using a tuned circuit that is in neither collector nor emitter but is linked by mutual coupling to both electrodes. Figure 7.1d shows a tuned-collector oscillator in which the tuned circuit is connected in series with the transistor, mutual coupling providing feedback as before. The resonant frequency in both circuits is $1/2\pi\sqrt{(LC)}$ where L and C represent the *total* effective inductance and capacitance of the circuit; it is adjusted by alteration of either inductor or capacitor, usually the capacitor because this is easier to alter.

Figure 7.2 shows the Hartley (parts a, b and c) and Colpitts (parts d and e) families of oscillator circuits. In the Hartley circuits shown all three versions are similarly connected to the tuned circuit: the coil tapping is connected to the cathode, directly in the shunt and inverted circuits and via the power supply in the series circuit; the $L1 + C$ junction is connected to the anode, directly in the series and shunt circuits and via the power supply in the inverted circuit; the $L2 + C$ junction is connected to the grid, directly in all three circuits. The Hartley circuit is rearranged (purely for explanatory purposes) in figure 7.2f showing that the anode-cathode voltage is connected across C and $L2$, and the grid–cathode voltage is across the coil $L2$. If $L2$ is small, the $L2 + C$ combination is mainly capacitive and the current in this section leads the applied voltage (that is, the anode–cathode voltage) by 90°. The voltage across $L2$ leads the current through it by approximately 90° so that the grid–cathode voltage across $L2$ will lead the anode–cathode voltage across $L2 + C$ by almost 180°. The tuned circuit being at resonance presents a resistive load to the amplifier stage. This gives a further 180° phase shift. The

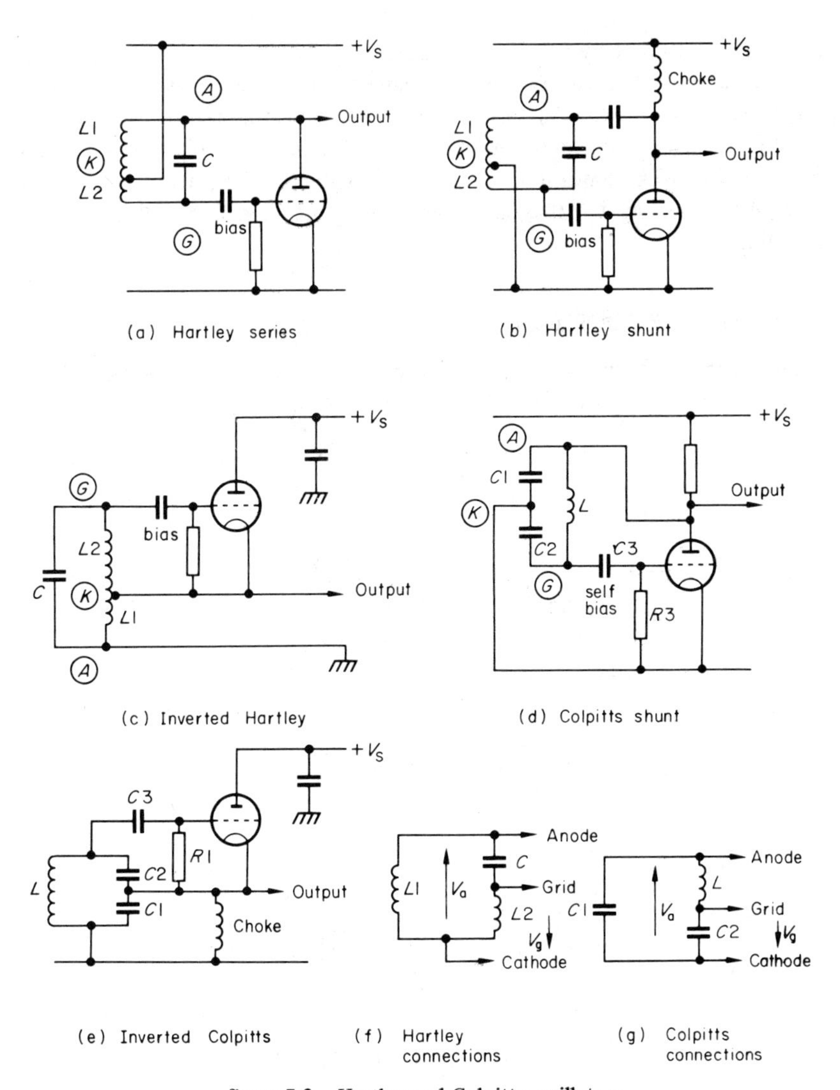

figure 7.2 Hartley and Colpitts oscillators

overall phase shift is thus almost 360° and conditions for positive feedback are satisfied. Feedback is adjusted by altering the ratio $L2/L1$. The oscillator frequency is $1/2\pi\sqrt{(LC)}$ where L is the combined inductance of $L1$ and $L2$.

The Colpitts oscillator shown in figure 7.2d and 7.2e has a tuned circuit with tapped capacitor, the effective connections in both circuits being those shown in figure 7.2g. Here, the grid–cathode voltage across $C2$ is phase shifted by almost 180° from the anode–cathode voltage as the voltage across

$C2$ lags the current by 90° and the current (which is that flowing in L and $C2$, a circuit branch which is mainly inductive) lags the applied anode–cathode voltage by nearly 90°. As with the Hartley circuit, the amplifier (being resistive-loaded) provides a further 180° phase shift, so the positive feedback condition is satisfied. With the Colpitts oscillator the resonant frequency is $1/2\pi\sqrt{(LC)}$ where C is the capacitance of $C1$ and $C2$ combined.

7.2. Feedback *RC* Oscillators

Phase shifting of one voltage with respect to another is possible using a resistor–capacitor network. For example, in a simple series *RC* circuit the

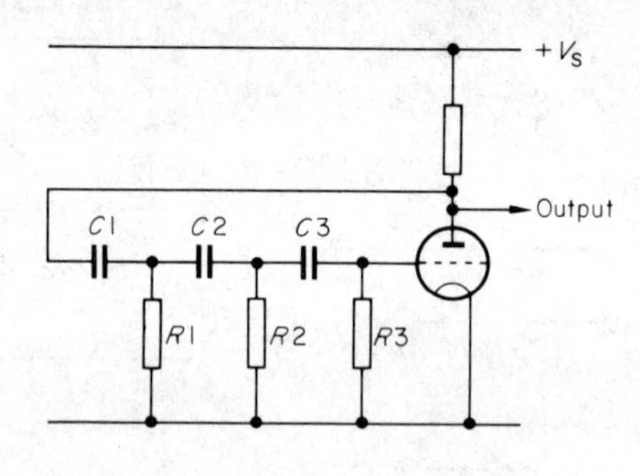

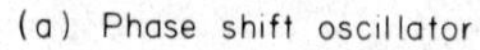

(a) Phase shift oscillator

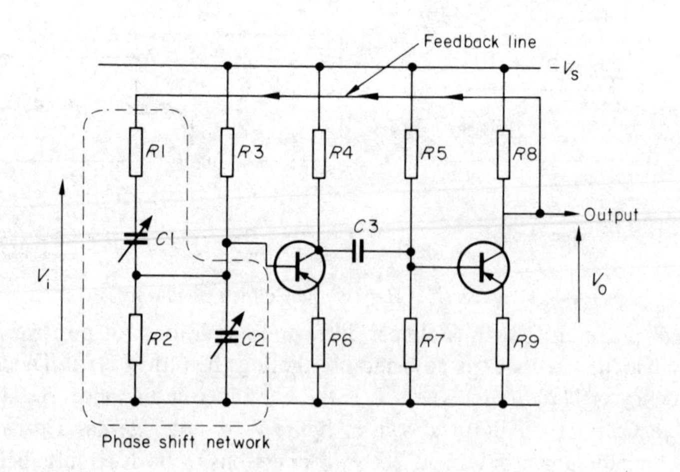

(b) Wien Bridge oscillator

figure 7.3 Feedback *RC* oscillators

voltage across the resistor will lead the applied voltage by an angle lying between 0° and 90° and determined by the values of R and C. (For $C = 0$, phase shift is 0°, the circuit being purely resistive; and for $R = 0$, phase shift is 90°, the circuit being purely capacitive.) An RC phase shift oscillator, often used in laboratory test instruments, is shown in figure 7.3a. Three RC networks in cascade provide a total phase shift lying between 90° and 180°, and the amplifier provides the other 180°. The feedback fraction is adjusted by altering the values of R and C and the resonant frequency is given approximately by $1/2\pi CR\sqrt{6}$.

A Wien bridge oscillator is shown in figure 7.3b. A Wien bridge is an electrical measurement device consisting of a specific arrangement of resistors and capacitors, and it may be used to determine component values. At one frequency determined by components $R1$ and $C1$ the phase shift between V_o and V_i in the circuit shown is zero (360°). No further phase shift is required so two resistive loaded amplifier stages are employed, thereby providing a total phase shift of 180° + 180° = 360° (or 0°). Feedback fraction is adjusted by altering component values and the resonant frequency is $1/2\pi C1R1$. The circuit is not suitable for adjustment over a wide range of values of output frequency. The output waveform of both the RC phaseshift oscillator and the Wien bridge circuit is sinusoidal and has a high degree of purity (that is, it has few harmonic components.)

7.3. Crystal Controlled Oscillators

Certain materials in crystalline form, particularly quartz, exhibit what is known as the piezo-electric effect (piezo means pressure). A pressure exerted on the faces of such a crystal sets up a small voltage. Alternatively, voltage applied across the crystal causes contraction or expansion of the crystal. If a small oscillating voltage is applied to the crystal, mechanical vibrations are set up at a frequency determined by the size, shape and cut of the crystal. These vibrations increase the voltage and maintain the oscillation at the given frequency with a high degree of stability. Crystals, in fact, behave like a highly selective tuned circuit with a very sharp frequency response curve. To improve frequency stability of oscillators, in particular those used in transmitting systems, it is common practice to employ crystal control.

A highly stable oscillator circuit used for frequency trebling is shown in figure 7.4. Here a crystal is used in an oscillator to set up a stable lower frequency and the third harmonic of this output is then selected by a suitably tuned circuit. In this way high frequency stability is obtained at frequencies higher than those normally controllable by crystals. Stability is further

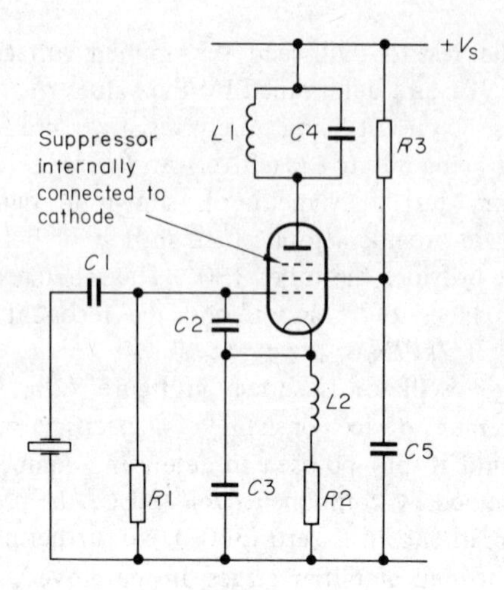

figure 7.4 Electron coupled oscillator (crystal controlled)

improved by a technique called *electron-coupling* (which is discussed below after the circuit description).

The oscillator is a Colpitts oscillator using the screen grid, control grid and cathode of the pentode as anode, grid and cathode of a triode. The crystal is shunted by the capacitors $C2$ and $C3$ tapped at the junction, which provides the required Colpitts arrangement and ensures the necessary phase shift. Components $C1$ and $R1$ provide automatic bias and $R2$ cathode bias (which, in the event of signal failure, prevents valve damage). The Colpitts circuit is of inverted form, the junction between the crystal and $C3$ being linked to the effective anode (the screen grid) via the power supply and screen biasing resistor $R3$. The output from the $C2 + C3$ junction is linked to the cathode which is loaded by choke $L2$.

An inverted Colpitts circuit normally provides the output from the cathode, but in this design the valve anode is loaded with a tuned circuit and the output is then taken from the anode. The actual output (anode) is coupled to the usual output (cathode) by the electron stream within the valve. In this way, the oscillator, which uses a triode equivalent for the generation of the signal, is coupled to the actual output using a pentode. As anode voltage of a pentode has little effect on anode current (examine the pentode characteristics, chapter 4), any change in anode voltage resulting from variation in supply (or caused by following stages), which would normally affect the oscillator output if a triode were used on its own, has a much reduced effect. Electron coupling thus improves stability. Further trebling stages may be added to give a stable output as many as 18 times the crystal frequency. Crystals may be used together with, or to replace, the tuned circuit in any of the *LC* oscillators so far considered with consequent improvement in frequency stability. Each crystal has one characteristic frequency that is determined by size and cut etc., and cannot be used at any other. Crystals are particularly useful for purposes where the oscillator frequency must remain constant; for example, in the generation of carrier waves in transmitting systems.

7.4. Negative-resistance Oscillators

As has been stated, a tuned circuit will oscillate of its own accord if a voltage is applied across it and then removed. During one half cycle the capacitor is charged and stores energy in the electric field between the plates. Once charged the capacitor discharges and the electric field collapses, transferring the energy to the inductor magnetic field which now begins to build up. The magnetic field reaches a maximum then begins to collapse, transferring the energy back to the capacitor. The process continues, generating a sinusoidally varying voltage as it does so. If it were possible to have a circuit with no resistance the oscillations would continue indefinitely. However, as the coil and connecting wire do have resistance, energy is lost at each transfer and the oscillations die away. The applied voltage–current graph for a resistor shows that if the voltage is increased the resistor current increases, and that the current falls if the applied voltage is reduced. In some electronic components (notably the tetrode valve and the tunnel diode) a rising voltage causes a falling current. This property is the opposite of normal resistance and (as previously indicated) is called 'negative resistance'. Negative resistance returns energy to the circuit; positive resistance dissipates it as heat.

If a tuned circuit or tuned circuit equivalent, such as a crystal, is connected to a device that exhibits negative-resistance, the natural oscillations of the tuned circuit may be maintained. Figure 7.5a shows a crystal oscillator using a tunnel diode to counteract the internal energy loss within the crystal. The frequency of the output is the natural frequency of oscillation of the crystal. An oscillator using a tetrode in a similar fashion is called a *dynatron-oscillator,* but as this circuit is not commonly used nowadays it is not illustrated.

Another form of negative resistance oscillator, though not using an obvious negative-resistance device, is the tuned-anode tuned-grid circuit (tuned-collector tuned-base when using transistors). In this circuit two tuned circuits are used, one between grid and earth and one between anode and supply. The circuits are NOT mutually coupled. Feedback occurs via the

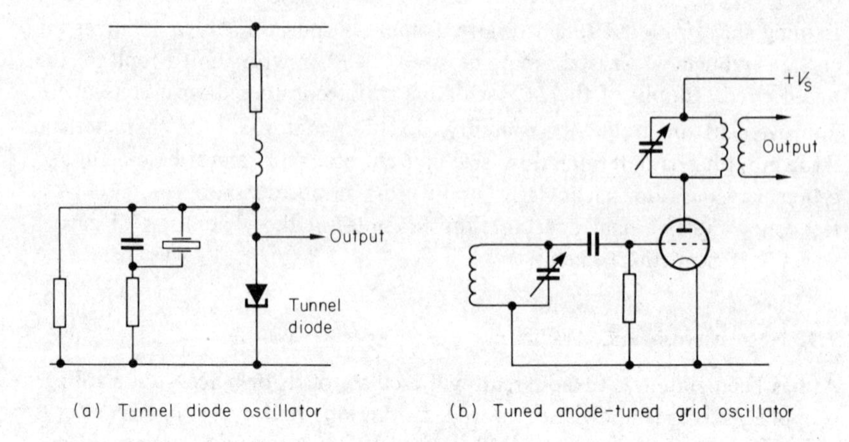

figure 7.5 Negative resistance oscillators

internal interelectrode capacitance between anode and grid and is called Miller feedback after its discoverer. Miller feedback has the same effect as shunting the grid tuned circuit by a component that may under appropriate circuit conditions become resistive, capacitive or inductive. If the tuned circuit in the anode has a slightly *higher* resonant frequency than the grid circuit, it is found that a negative resistance effect is felt at the grid and the natural oscillations of this circuit at the resonant frequency are thereby maintained. (See figure 7.5b.)

7.5. High-frequency Oscillators

Oscillators for use in the very-high and ultra-high frequency bands (VHF and UHF) operate on the same principle as oscillators used in the lower frequency bands. However, the components may differ radically in appearance, or even the internal capacitance or inductance of the valve or transistor may be used leaving no actual component visible.

Frequency multiplier circuits, one of which is illustrated in figure 7.4, may be used for high frequencies up into the VHF band. In this circuit, as was explained earlier, the third harmonic of the crystal controlled lower frequency is picked out by a suitably tuned *LC* circuit. Further multiplier stages may follow to give a highly stable output at many times the original frequency.

An ultra-audion oscillator circuit is shown in figure 7.6. The principle here is that of the Colpitts oscillator but the capacitors are not visible because it is the anode–cathode and grid–cathode capacitances (which are too small to be

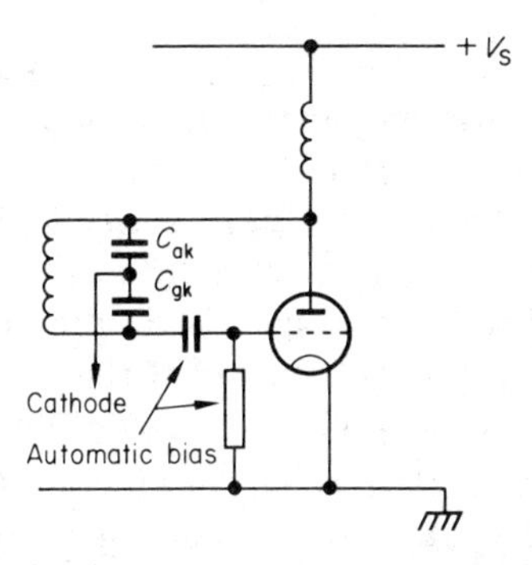

figure 7.6 Ultra-audion oscillator (C_{ak}, C_{gk} are valve internal capacitances and are not actually visible in circuit)

useful at lower frequencies) that are put to use to give an output frequency well into the UHF band. The oscillator frequency may be adjusted by the addition of a small variable capacitor connected in parallel with the coil.

All conducting wires have inductance and, in pairs, have capacitance between wires, the wires acting as capacitor plates. At all but very high frequencies, the values of this inductance and capacitance are so small that their effect is negligible. However, in the VHF band and above, the inductance and capacitance forming part of a *transmission-line* (normally used for transferring high frequency signals from transmitter to aerial or aerial to receiver in telecommunication systems) are extremely significant; as, provided the line length is of the correct size relative to the wavelength appropriate to the oscillator frequency, a transmission line section may be used as an effective tuned circuit to replace the more conventional

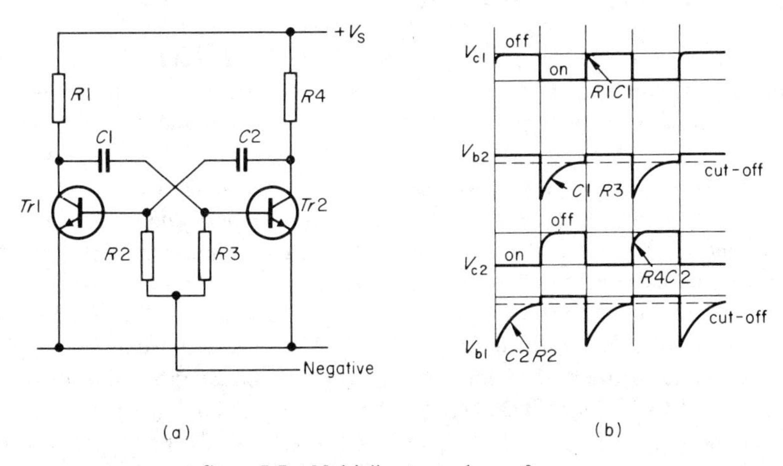

figure 7.7 Multivibrator and waveforms

components. Such sections, called *lecher-bars* (or hairpin tanks in the USA), are commonly used in radar and other UHF applications. Transmission lines are dealt with in more detail in chapter 10.

At higher frequencies in the UHF band, *cavity resonators* are used. These look like metal boxes of spherical, square or rectangular shape. In all cases they can be considered to be made up of numerous quarter-wave lengths of transmission line joined together to form continuous surfaces. Each resonator is effective at the one frequency that is determined by the size and shape of the enclosed cavity. Other devices employed at UHF include magnetrons, klystrons, Gunn diodes (the solid-state equivalent of the klystron), travelling-wave tubes, masers and lasers. The basic principles of operation of these devices were considered in chapter 4.

7.6. Relaxation Oscillators

One of the most widely used forms of relaxation oscillator is the astable multivibrator, a transistor form of which is shown in figure 7.7a. This type of circuit using two switching transistors (or valves) is the basis of three commonly used circuits: the monostable, or one-shot circuit; the bistable circuit, or flip-flop; and the form shown here, the astable circuit. The word monostable or bistable describes the number of available stable states in which the circuit will stay until an external trigger pulse is applied. (The mono- and bi-stable circuits are described in chapter 9.) The astable circuit has no stable states and switching between states proceeds continuously. The circuit is thus free-running and may be used as a pulse generator or square-wave oscillator. Astable multivibrators are often used as pulse-train generators in logic circuit applications, especially computers.

In one of the two states *Tr*1 is conducting and *Tr*2 is cut off, and in the other *Tr*2 is conducting and *Tr*1 is cut off. The waveforms are illustrated in figure 7.7b. Starting with *Tr*1 off and *Tr*2 on and just before switching takes place, *Tr*1 collector voltage $Vc1$ is high (at the supply voltage); *Tr*2 collector voltage $Vc2$ is low; *Tr*1 base voltage $Vb1$ is rising exponentially (Note: An exponentially changing voltage is one in which the *rate of change* is falling as the change progresses. A capacitor always charges and discharges to an exponential pattern.) towards the cut-off voltage; and *Tr*2 base voltage $Vb2$ is at a positive voltage sufficient to cause full conduction in *Tr*2. When $Vb1$ reaches cut-off, *Tr*1 starts to conduct; $Vc1$ begins to fall from the supply voltage towards the saturation voltage for full conduction; this drop is transmitted to $Vb2$ via $C1$ and (as $C1$ cannot change its p.d. instantaneously) $Vb2$ falls sharply, cutting off *Tr*2. $Vc2$ now rises towards the supply voltage and this rise is transmitted via $C2$ to $Vb1$, which increases slightly above

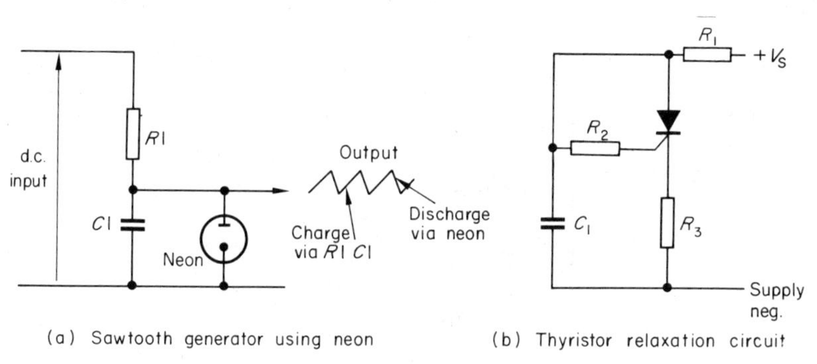

figure 7.8 Relaxation oscillators

cut-off, thus driving *Tr*1 into full conduction. *Tr*1 is now fully conducting, and *Tr*2 is cut-off. *V*b2, which was driven to a negative low by the change in *V*c1, now starts to rise as *C*1 begins to charge via *R*3. Eventually *V*b2 reaches cut-off thereby initiating *Tr*2 switching on and *Tr*1 switching off. As the circuit has now been restored to its original condition, it is clear that the entire process is repetitive. The exponential change of *V*b1, with which the description of operation commenced, is due to *C*2 charging via *R*2. Other exponential changes in the waveform are the rise in *V*c1 and *V*c2 to the supply voltage when their respective transistors are cut off. *V*c1 rises exponentially due to *C*1 charging via *R*1 and *V*c2 rises exponentially due to *C*2 charging via *R*4. The width of the pulses taken from either collector is determined by time constants *C*1*R*3 and *C*2*R*2.

Figure 7.8a and b shows two other forms of relaxation oscillation using a single active device. Part (a) shows a simple neon saw-tooth generator in which the neon lamp is the active device. A neon lamp is a small version of a gas filled diode which glows when the gas is ionised. When a d.c. supply is applied across the series combination of *R*1 and *C*1, the capacitor begins to change exponentially with time constant *R*1*C*1. (Note: The time constant of a circuit containing resistance *R* ohms and capacitance *C* farads is given by *CR* seconds. It is the time for the capacitor p.d. to change by 0.63 of the maximum value.) If *R*1*C*1 is large then the output voltage rises approximately linearly at first. When the output voltage reaches a certain value (known as *the striking potential*) the neon lamp conducts and *C*1 discharges rapidly. The process now repeats. The width of the saw-tooth is determined by the time constant *R*1*C*1. Other saw-tooth generators are available based on the same principle but using thyratrons (gas filled triodes) or the solid-state equivalent (thyristors). A thyratron or thyristor conducts when a suitable trigger pulse is applied to the control electrode, or 'gate', which then loses control. Conduction is stopped by reducing the anode voltage to below the 'holding

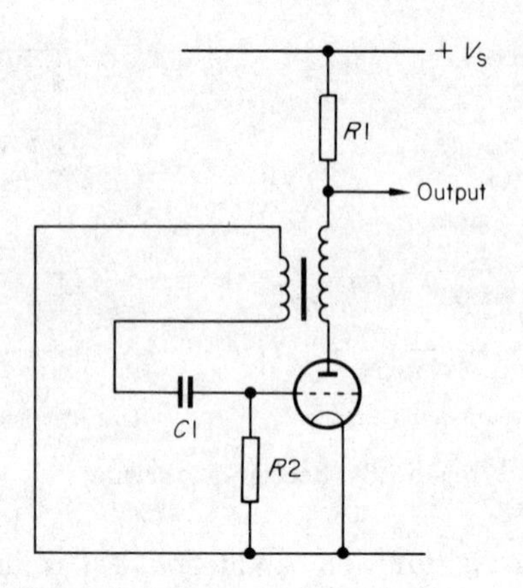

figure 7.9 Simple blocking oscillator circuit

voltage'. One such low frequency generator using a thyristor is shown in figure 7.8b. On application of a direct voltage where shown, capacitor $C1$ charges via resistor $R1$. Eventually, a point is reached where the voltage applied to the gate via $R2$ is sufficient to trigger the thyristor and $C1$ discharges. When the thyristor anode voltage falls below the holding value the thyristor switches off and the cycle is repeated.

The final circuit to be considered in this section is the blocking oscillator, one form of which is shown in figure 7.9. This circuit resembles the *LC* oscillator shown in figure 7.2a and the circuit as shown will oscillate if the appropriate conditions are set up. However, the feedback fraction, the internal resistance of transformer, and the values chosen for the bias circuit components $R2$ and $C1$, ensure that only one or two oscillations occur before the circuit is blocked (that is, cut off) by the automatic bias produced by the first one or two cycles. There is then a period of waiting while $C1$ discharges via $R2$ and the grid coil. After $C1$ has discharged oscillations build up again until they are again blocked. The circuit produces a regular pulse output, the *repetition rate* being determined among other things by the bias circuit components and by the design and construction of the transformer.

8 Mixing and detection

In a tuned radio frequency receiver the modulated carrier wave is received by a tuned input stage then amplified by further stages before the intelligence is extracted. All r.f. stages handling the signal must be tunable: that is, it must be possible to change the frequency at which maximum gain occurs. This requirement reduces the number of amplifying stages because it is difficult to arrange satisfactory tuning of more than two or three stages. The shortage of stages thus renders the receiver insensitive to small amplitude signals. In addition, a TRF receiver has reduced *selectivity* (that is, the ability to select a particular carrier frequency among many) because the fewer the number of stages of amplification the wider is the band width of the overall frequency response curve. A *supersonic heterodyne* receiver (or *superhet*) overcomes these problems by the use of a *mixer stage* at the input. The mixer stage combines the incoming frequency and a locally generated oscillator frequency to give an *intermediate frequency* that carries the same modulation as was carried by the original carrier. When the receiver is tuned, both the input tuned circuit and the oscillator are tuned together, so that the oscillator frequency is separated from the incoming frequency by a constant amount, and the intermediate frequency is thereby constant regardless of the frequency of the incoming carrier. This means that several further stages of amplification to improve sensitivity and selectivity can be employed and the tuning of these stages, when once set, can remain fixed. The superhet is thus far superior to the TRF receiver in performance. Sometimes, in fact, steps have to be taken to *reduce* selectivity because the resultant overall bandwidth will not accommodate the sidebands of the desired carrier. The special coupling techniques used to do this were considered in chapter 6.

Detection is the process of extracting the intelligence from the incoming signal. The circuits used depend on the type of modulation that was used to impress the intelligence signal on the carrier. Frequency modulated signal detectors are often called *discriminators.*

8.1. Mixer Circuits

The mixer circuit combines two input frequencies to give a composite signal containing a number of frequencies. The required intermediate frequency is then extracted from the composite signal. The process is similar to the modulation process that takes place at the transmitter and the extraction of the i.f. at the mixer stage is a similar process to the extraction of the a.f. later in the receiver (that is, detection). This is why the mixer stage is often referred to as the first detector to distinguish it from the second detector which extracts the audio intelligence.

There are two types of mixing: *additive mixing,* which combines the two input signals at the same electrode of the valve or transistor; and *multiplicative mixing,* in which the inputs are applied to separate electrodes. One of the inputs is from the local oscillator, which is tuned at the same time as the input tuned circuit, and this oscillator may be either a separate circuit or part of the mixer. Where it is part of the mixer, the overall circuit is often referred to as a *frequency converter.*

Figure 8.1 shows two forms of additive mixer. In 8.1a (the valve version) the oscillator is separate and is not shown. In 8.1b (a Meissner type) the oscillator is part of the mixer circuit and the oscillator signal is generated within the transistor to the base of which the input carrier is fed. The valve or transistor has a tuned circuit as the load, the resonant frequency being preset to the i.f. In figure 8.1b the oscillator frequency is adjusted by *C*5 which would be *ganged* with the capacitor of the input tuned circuit so that one adjustment alters both capacitors. The oscillator frequency must always be separated from the input carrier frequency by an amount equal to the i.f., and this implies that the oscillator and input circuit must always *track* correctly. To improve tracking at low frequencies a series capacitor *C*3, called a *padder,* is inserted; and for tracking at high frequencies a shunt capacitor *C*4, called a *trimmer,* is inserted. Both capacitors, which are used in all the tunable circuits at the input side of the receiver, are preset when the set is *aligned.* The user then has to vary only one control to change the main variable capacitor which comprises both *C*5 and the capacitors that tune the input circuits. An r.f. stage is not usually included before the mixer stage in domestic receivers (for reasons of economy) but where one is used it must be tunable, and its tuned circuit must also contain appropriate padders and trimmers. These comments on alignment and tracking apply equally to valve receivers.

For multiplicative mixing special multi-grid valves are used, and these include the hexode and the heptode (pentagrid). In the *hexode* the signals are applied to grids 1 and 3, grids 2 and 4 being connected together and treated

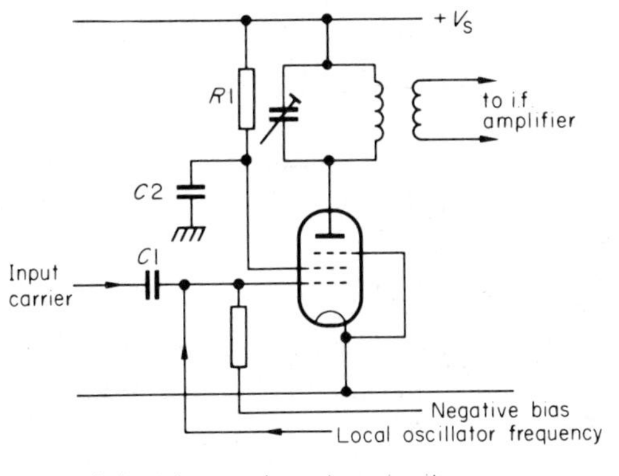

(a) Using a valve mixer circuit

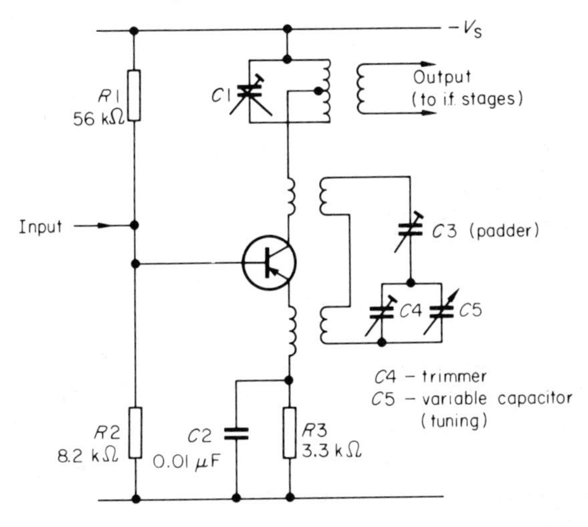

(b) Using a transistor converter circuit

figure 8.1 Additive mixing

as the screen grid of a pentode. Figure 8.2 shows a mixer circuit using a valve containing the hexode-mixer and the oscillator-triode in one envelope. The oscillator shown is a shunt-connected tuned-anode oscillator, the grid of the triode being connected internally to grid 3 of the hexode. The valves share a common cathode. In the circuit shown, *R*1 and *C*1 provide decoupled screen voltage for grids 2 and 4; *R*2 and *C*2 provide cathode bias; *C*3 and *R*4 provide automatic bias for the oscillator; *C*4 blocks the d.c. level at the triode anode from the oscillator tuned circuit; *C*5 and *C*6 are used for alignment as already described; and *C*7 is the variable capacitor. Notice that band-pass coupling, as

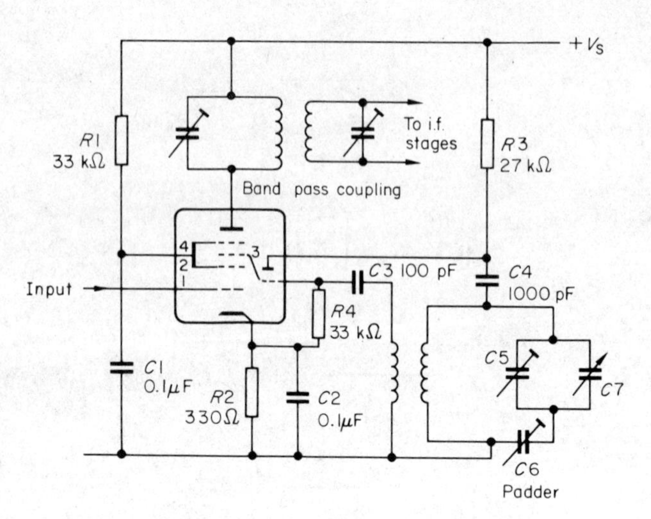

figure 8.2 Triode hexode mixing (multiplicative)

described in chapter 6, is used to increase the bandwidth to anticipate the inevitable reduction caused by the i.f. stages.

8.2. AM Detectors

The shape of a typical AM carrier is shown in figure 8.3a. If such a signal is fed to a circuit containing a diode, capacitor and resistor, the waveform across the resistor will be that shown in figure 8.3b. The circuit may be arranged as a series circuit (figure 8.3c) or a shunt circuit (figure 8.3d). In both circuits the diode conducts on positive half-cycles of the carrier and cuts off on negative half-cycles. The capacitor charges during the period when the signal voltage is greater than the capacitor voltage and discharges when the signal voltage falls below the capacitor voltage. In the circuits shown, provided the time constant $C1R1$ is the correct value, the output voltage will follow the shape of the audio modulating signal quite faithfully. If $C1R1$ time constant is too long, the output waveform cannot follow the troughs in the audio wave because the capacitor cannot discharge quickly enough, so that some fidelity is lost. If $C1R1$ time constant is too short, the output waveform carries a pronounced radio-frequency ripple.

The rectified waveform of figure 8.3b is made up of three components: the required audio output; a radio-frequency ripple; and a d.c. level about which the audio signal varies. The surplus components are removed by a filter circuit which may be of the form shown attached to the series detector in

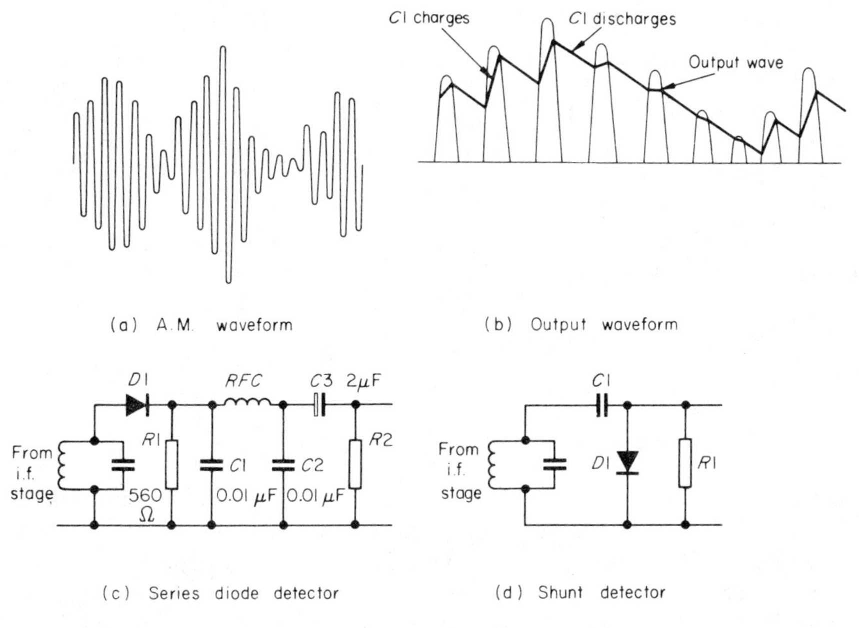

figure 8.3 Detection

figure 8.3c. The circuit consists of a 'potential-divider' arrangement comprising a radio-frequency choke (*RFC*) and capacitor *C*2; with a further series circuit, consisting of *C*3 and the load *R*2, connected across *C*2. Capacitor *C*2 is of such a value that it presents a low reactance to r.f. and fairly high reactance to a.f. Thus the a.f. component appears across *C*2. The r.f. choke and *C*2 may be compared to a mains choke and output capacitor in a power supply filter circuit. Capacitor *C*3 blocks the d.c. level and presents a low reactance to a.f. so that the desired audio frequency output appears across the load *R*2.

When comparing detector circuits the important characteristics to be taken into consideration are the effects that each circuit has on the previous r.f. (or i.f.) stages, particularly the effect on selectivity and sensitivity. As has been seen, damping a tuned circuit by shunting it with a resistor flattens the impedance-frequency curve and thus (if the tuned circuit is the load of an amplifier) the gain-frequency response curve. Flattening this curve reduces gain (which reduces sensitivity) and widens the bandwidth (which reduces selectivity). The important question therefore is whether or not the detector loads the tuned circuit. Diode detectors have the disadvantage of drawing current from the input circuit so that their effective input resistance is low.

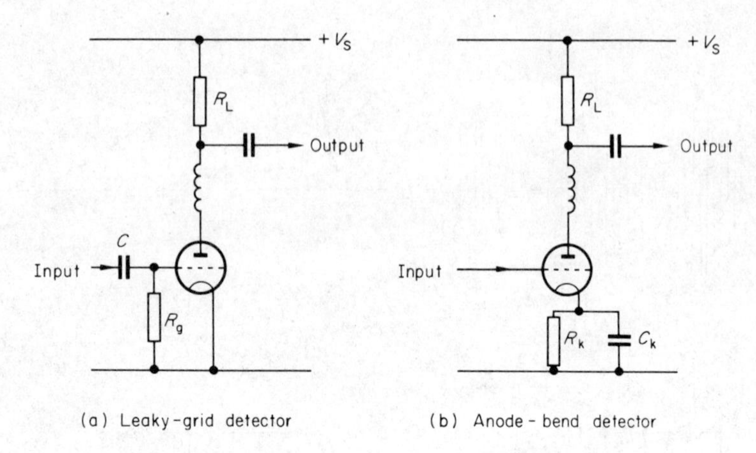

figure 8.4 Standard amplitude-modulation detectors

Selectivity and sensitivity are thus reduced. Also, since the diode offers no amplification, the detector circuit cannot compensate for the reduced sensitivity. Compensation can be introduced to some extent, however, by increasing the gain of previous stages. Reproduction of the audio signal is good with diode detection except that with semiconductor diodes and weak signals the reverse *IV* curve reduces the effectiveness of the initial rectification.

Circuits that detect and amplify include the 'leaky-grid' detector and the 'anode-bend' detector. Both these circuits are possible with field-effect transistors (although, of course, the names would be changed). A leaky-grid detector is shown in figure 8.4a. Bias is provided by an auto-bias *RC* circuit as shown and depends upon the input signal strength. Detection takes place between grid and cathode (which behaves as a diode) but triode action produces amplification of the signal, the output being taken from the anode. Due to grid current, which establishes bias, the circuit loads the previous stage, reducing selectivity and sensitivity. The reduction in sensitivity is compensated by the amplification of the detector. Fidelity of a leaky-grid circuit is not as good as that provided by diode detection.

An 'anode-bend' circuit is shown in figure 8.4b. Here the valve or transistor is biased almost to cut-off so that the lower part of the dynamic transfer curve is used, the grid-cathode combination detecting as before. Bias is provided either by a cathode-bias circuit (using high valued resistors because of the low quiescent current) or by an external source, so that sensitivity and selectivity are not reduced by grid current. Anode-bend detection is commonly used in frequency-changers, the circuit of figure 8.1a (which obtains its bias externally) being an example. The fidelity of an anode-bend detector is better than that of a leaky-grid detector; however, a reduced gain at the lower end of the transfer curve reduces the detector sensitivity for weak signals.

8.3. FM Detectors (Discriminators)

In FM transmission the amplitude of the carrier signal remains constant and the frequency changes according to the audio signal variation. A circuit to detect FM signals must therefore be a frequency-sensitive circuit that produces an output voltage that depends upon the input frequency variation. Circuits commonly used in FM detectors (or discriminators) are series or parallel tuned circuits that behave resistively at the resonant frequency (equal to $1/2\pi\sqrt{(LC)}$ where L and C are the values of inductance and capacitance) and either inductively or capacitively as the supply frequency moves away from the resonant frequency. Three common circuits are given: the stagger-tuned discriminator; the Foster-Seeley discriminator; and the ratio detector. Slope detection is also considered. The discriminator circuits are shown in figure 8.5.

Figure 8.5a shows a stagger tuned discriminator. Here two tuned circuits are transformer coupled to the final stage of the i.f. amplifier. One of the circuits is resonant at a frequency above the centre carrier frequency, the other at a frequency below it. The voltage developed across each circuit is rectified and the overall output voltage is the difference between these two rectified outputs. At the centre frequency one tuned circuit is inductive in nature, the other capacitive. If the separation between centre frequency and resonant frequency is the same for both circuits two equal and opposite voltages are developed, one across each circuit, and the output is then zero. At a carrier frequency above centre the voltage developed across one tuned circuit is larger than that across the other because one circuit is further from resonance. The output is no longer zero but is the difference between the individual outputs (for they are approximately in antiphase). When the carrier frequency deviates in the opposite direction (that is, below centre) the other tuned circuit produces a larger voltage, so that the difference between the two outputs is again not zero but has a value in antiphase to the previous voltage. Thus the discriminator produces a varying voltage that is 'positive' or 'negative' as determined by the carrier frequency deviation. This circuit is sensitive to amplitude variations and an amplitude limiter circuit (see article 8.4) is normally incorporated prior to the discriminator.

Figure 8.5b shows a Foster-Seeley discriminator (named after its designers). This circuit uses only one tuned circuit, resonant at the centre-frequency of the FM carrier. The circuit is transformer fed from the i.f. stage and thus behaves as a *series resonant circuit.* A series resonant circuit

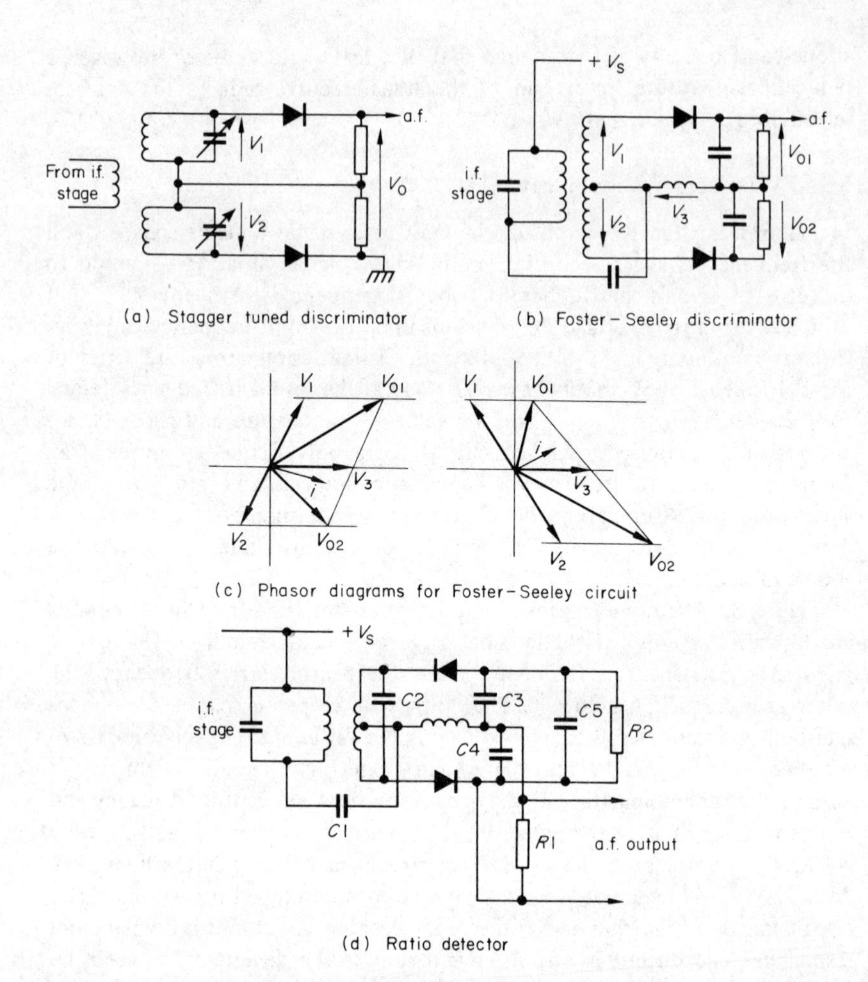

figure 8.5 **Standard frequency-modulation detectors**

is *inductive* at frequencies *above* resonance and *capacitive* at frequencies *below* resonance. In the circuit shown voltage V_3 is equal to the voltage across the primary side of the input circuit (the last stage of the i.f. amplifier) and the secondary side is centre-tapped to produce two equal antiphase voltages, V_1 and V_2. Output voltage V_{01} is the sum of V_1 and V_3 and V_{02} is the sum of V_2 and V_3 (V_3 being used as the common reference voltage). The overall output is the sum of V_{01} and V_{02}. At frequencies above the centre frequency the circuit is inductive and the circuit current i lags the supply voltage V_3. Voltages V_1 and V_2 are in quadrature with i (Note: this means that there is a 90° phase shift between the voltage and the current) and in antiphase with each other, as shown in the phasor diagram of figure 8.5c. V_1 and V_3 are summed vectorially to give V_{01}, V_2 and V_3 to give V_{02}. As can be seen, V_{01} is greater than V_{02} and the overall output is not zero and is shifted in phase from V_3. At frequencies below centre, the situation is as shown in the second part of the figure, with the current leading V_3, for the circuit is now capacitive. V_1 and V_2 are again in quadrature with i and in antiphase with each other, but their new phase position makes V_{02} (the sum of V_2 and V_3) greater than V_{01} (the sum of V_1 and V_3). The overall output depends upon the relative phase and magnitude of V_1 and V_2, upon the position of i, and thus upon the frequency deviation of the input. As with the stagger tuned circuit this discriminator is also sensitive to amplitude variation and a limiter circuit must be used.

The ratio detector shown in part (d) of figure 8.5 has a similar circuit to the previous discriminator with one important difference. Here, one diode is reversed and the overall output is the sum of the individual output voltages. The inclusion of $C5$ across $R2$ holds the voltage across it relatively constant and amplitude variations are not as effective in changing the output (that is, the circuit acts as its own limiter). The output, which is taken from across one of the capacitors, varies in amplitude according to the input frequency deviation, as with the previous circuits.

Narrow band FM transmissions can be detected using an AM detector circuit if the receiver is tuned slightly off centre. The gain of an AM receiver is frequency sensitive and so variation in input frequency produces an output varying in amplitude. The method is called *slope detection,* because the slope of the AM receiver gain/frequency curve is used. The method is limited to narrow band transmission because a linear part of the curve is required for low distortion.

8.4. Amplitude Limiters

Limiter circuits used to maintain constant amplitude in FM and other circuits may take the form of low gain r.f. amplifiers or may use a combination of diode clipping circuits.

The circuit of figure 8.6a shows a low-gain pentode r.f. amplifier. Similar transistor circuits are equally possible. The anode and screen grid voltages are reduced to lower values than normal (anode limiting) thereby reducing gain, and grid-limiting is achieved by using a *grid-leak* resistor $R4$. The grid bias set up by this resistor and capacitor $C1$ increases if the input signal increases and thus moves the operating point so that the amplifier gain is modified. Two or more limiter stages may be used in cascade to improve performance.

A diode limiter consisting of two diode clipping circuits is shown in figure

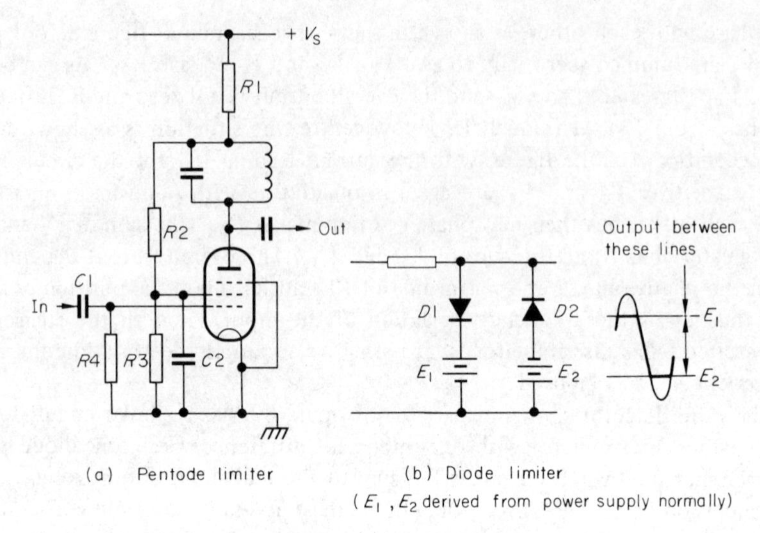

figure 8.6 Limiters

8.6b. During positive variation of the input signal diode $D2$ is cut off all the time and diode $D1$ remains cut off until the signal voltage reaches a level equal to $E1$. The output voltage thus follows the input voltage to this level. When level $E1$ is reached diode $D1$ conducts putting an approximate short circuit across the output. The output voltage now remains at $E1$ until the input voltage again reaches this level when $D1$ will cut off once more. The output positive half cycle thus looks as in the diagram alongside the circuit in figure 8.6b.

On negative half cycles of input voltage diode $D1$ is cut off throughout and $D2$ is cut off until the input voltage reaches a level equal to $E2$. At that point $D2$ conducts and as before the output voltage remains constant, only on this occasion at level $E2$. The circuit has thus 'sliced' a section of the input signal. The sine wave form is obviously changed by this technique but this is unimportant in a limiter that feeds a discriminator, because it is frequency that is the most important characteristic.

8.5. Automatic Gain Control

Radio signals are subject to changes in strength (fluctuations in amplitude) which vary from instant to instant in an unpredictable manner. Accordingly, most receivers (especially receivers of AM signals where the variation in signal strength is more important) include some form of *automatic gain control* (AGC) circuit that will combat such *fading* of the signal and thereby maintain a relatively constant output.

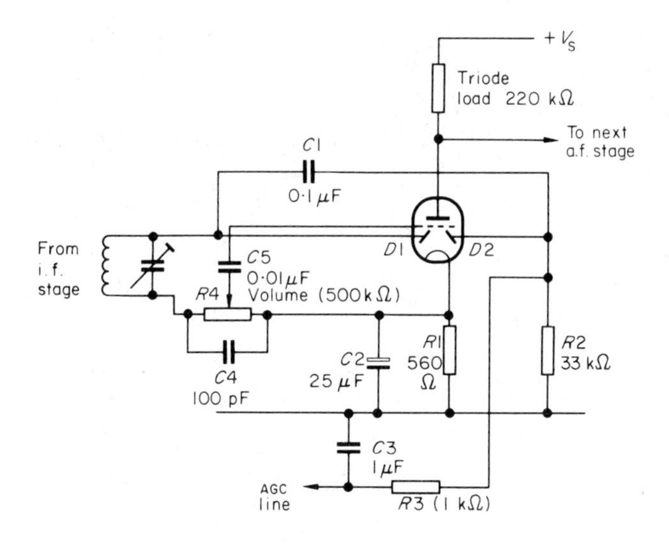

(a) Double diode triode AGC circuit

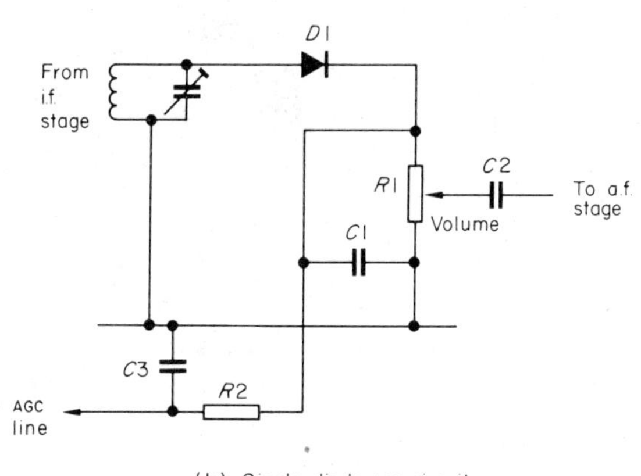

(b) Single diode AGC circuit

figure 8.7 Automatic gain control derivation

It was stated earlier in the chapter that an amplitude-modulated signal after rectification contains an a.f. component, an r.f. component, and a d.c. level proportional to the signal strength. The a.f. detector contains a filter circuit to remove the d.c. level and r.f. component. If AGC is required the a.f. and i.f. components are removed and the d.c. level is extracted and fed back to one of the r.f. amplifier stages. If the signal strength and thus the d.c. level falls, the r.f. amplifier gain is increased to compensate for the reduced signal. Similarly, if the signal strength increases, the AGC voltage is used to reduce

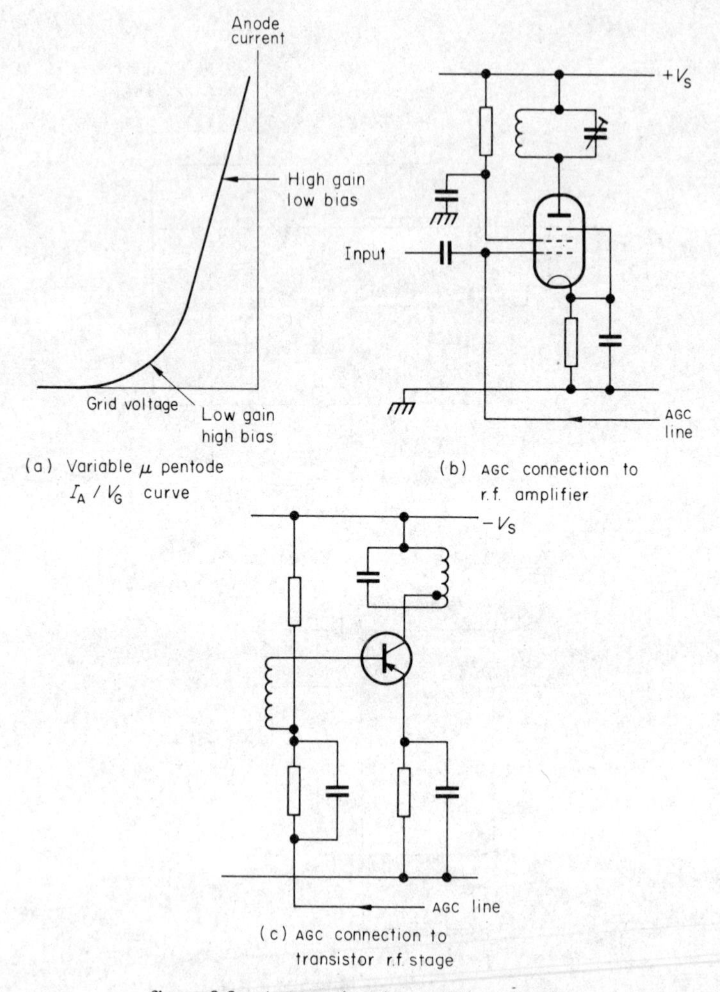

figure 8.8 Automatic gain control application

the gain of the r.f. amplifier. The output at the speaker then remains fairly constant.

Figure 8.7a shows a valve AGC circuit using a double-diode-triode. This valve, manufactured specially for this purpose, contains three valve structures in one envelope; two diodes and a triode sharing a common filament and cathode. One diode is used as an a.f. detector; its associated circuit containing a d.c. blocking capacitor *C*5, and an r.f. filter *C*4; and the a.f. being fed to the grid of the triode, which then acts as the first stage of the a.f. amplifier. *R*4 in this circuit is the audio volume control. The second diode is a shunt detector with $R2$ as load. $R3$ and $C3$ form an a.f. filter circuit and the AGC line carries only the d.c. level.

Figure 8.7b shows a single detector supplying both d.c. level to the AGC line and a.f. to the a.f. amplifier. Here $C1$ is an r.f. filter, $C2$ a d.c. blocking capacitor, and $R1$ the volume control. $R2$ and $C3$ form the a.f. filter. This circuit is commonly used with transistor circuits.

The simple AGC shown acts at all levels of signal. If it is not desirable to have AGC at low signal strengths, because the gain is held down by the AGC voltage, *delayed* AGC may be used. Here the AGC diode is biased at some preset voltage so that it will not conduct, and thus detect, until the signal reaches a certain strength. Some radio systems include an amplifier in the AGC line to increase the AGC voltage and the circuit is then termed an *amplified* AGC circuit.

The AGC voltage is fed back to an r.f. stage of the receiver. For a valve system, a special valve called a *variable-μ* pentode may be employed. With this device the I_A/V_G curve is non-linear and the gain, which depends on g_m and thus on the slope of this curve, varies with valve bias. The AGC voltage is fed to either the grid or cathode circuit in such a way that an increased AGC voltage reduces the gain by increasing the bias (see figure 8.8). Transistor gain generally depends on bias and on operating voltages, so the AGC voltage may also be used here to control gain as shown in figure 8.8c.

9 Switching and counting

Electrical or electronic switching is the act of controlling the passage of electrical power, or of signals carrying intelligence, from one subunit to another. There are many forms of switching, one or more of which can be applied to any switching device or circuit. These forms include remote or direct switching, manual or automatic switching, high speed or slow speed switching. The simplest and best known switch is the kind that is used for controlling mains electricity not only in low power domestic circuits but also in electronic systems such as oscillators, oscilloscopes, radios, televisions etc. This switch, which controls one supply line (the live wire) is a manually operated, slow speed, direct switch and is a mechanical device in which contacts move. For circuits in which voltage and current levels are higher and therefore offer a greater risk to the operator, remote switching is used. In remote switches a low power circuit controls a high power circuit, the low power circuit being operated either automatically or manually. Electromagnetic relays are an example of slow speed remote switches, the slowness being due to the presence of moving parts; thyratrons and thyristors are examples of remote switches operating at higher speeds due to the absence of moving parts. Very high speeds are achieved using specially constructed solid state devices made in integrated-circuit form. High speed switching and counting techniques are used in computing and in a variety of processes (such as industrial control) where the frequent and rapid measurement and monitoring of variable quantities is necessary.

9.1. Electromagnetic Relays

There is a large number of types of electromagnetic relay, ranging from miniature low-power devices, to very large high-power devices. All of them, however, contain moving contacts and are therefore slow speed switches. Two of the most widely used types are shown in figure 9.1. Part (a) of the figure

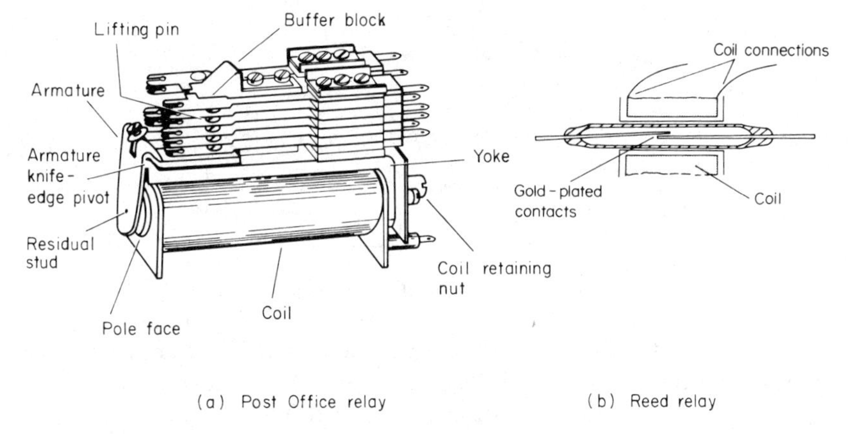

figure 9.1 Electromagnetic relays

shows a standard Post Office 3000-type relay extensively used in telephone systems and part (b) shows a reed relay. The Post Office relay consists basically of a moving armature which opens or closes a set of contacts. The armature movement is controlled by the magnetic field set up by a solenoid. The contact arrangement may consist of a single one-pole switch with normally-open, normally-closed, or changeover contacts; or may comprise many such switches each having normally-open, normally-closed, or changeover contacts, as required. (Note: a changeover contact unit is one that opens one circuit and closes another when the armature is attracted to the coil core.) Post Office relays are both simple and reliable in operation and versatile in the control facilities available. They are, however, fairly large and operate at low switching speeds. In addition, as there are moving parts, their reliability (though of a high order) is not as high as that of solid-state devices. The reed relay is an improvement of the conventional P.O. relay in that its contacts are in a sealed container, and the magnetising force required is much smaller. However, the individual switch contact arrangement is more restricted and, again, reed relays operate at relatively slow speeds. In the example shown, when current flows in the surrounding coil the reeds are magnetised and attract one another, thereby closing the circuit controlled by the switch. Multi-reed relays are available in which one coil controls a number of reeds similar in construction to that shown. Generally, reed relays are smaller than P.O. relays.

9.2. Electronic Power Switches

The two best known electronic devices for switching at relatively high power

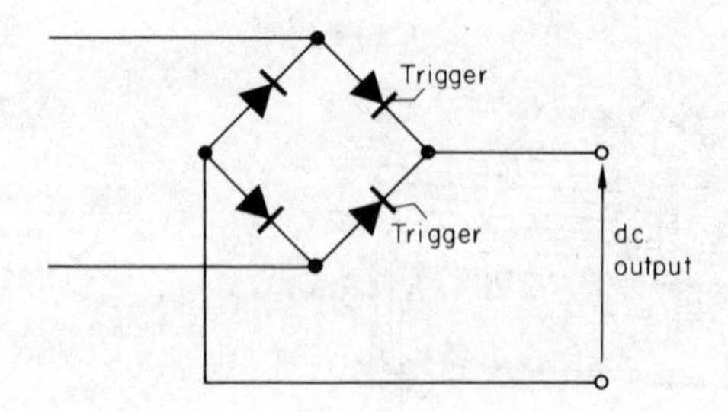

figure 9.2 Thyristor switching

levels are the thyratron and the thyristor. The thyratron is a gas-filled triode in which conduction is controlled by the grid (gate) voltage. Once conduction commences the grid has no further control, switching-off being effected by reducing the anode–cathode voltage. The thyristor is the solid-state equivalent of the thyratron and has similar properties and improved reliability. Both thyratrons and thyristors are widely used in industrial control circuits for controlling relatively high power levels. In the example shown in figure 9.2, low-power control circuits switch current to the thyristor gates; thus actuating the thyristors and thereby applying high-power levels to the load. One or both thyristors may be switched to provide half or full wave rectification. Thyratrons and thyristors operate at faster speeds than electro-mechanical switches but not as fast as the lower-power logic gates that will be considered in due course.

9.3. Gating

Gating and gate are terms (meaning switching and switch respectively) applied to the control of electronic signals carrying intelligence rather than to the control of electrical power. Gating circuits using valves or transistors operate at high speeds and are used for purposes ranging from selective signal control in television and telemetry systems to sophisticated calculating and decision making logic circuitry in computer and process control systems.

A gating circuit used for signal selection is shown in figure 9.3a. The signal, which may (as in figure 9.3b) contain pulses of various heights or widths, is applied to the control grid of a pentode that is held cut off by a sufficiently large negative voltage applied to the screen grid. To amplify the pulses marked 'A' in the incoming signal, a series of positive pulses at a voltage level sufficient to overcome the screen cut-off bias is applied to the screen grid. These *gating pulses* are synchronised with the 'A' pulses so that the valve is switched only during the pulses to be amplified. The valve is cut off at all other times. The output is an inverted, amplified version of the 'A' portion of

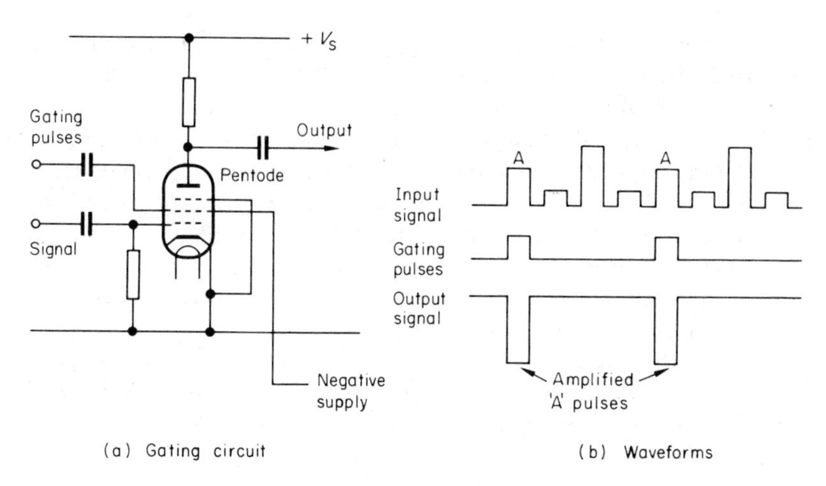

figure 9.3 Gating

the input. Devices used to gate signals in this manner are usually termed *logic gates* and, as there is a wide variety available, the next article deals solely with these items.

9.4. Logic Circuits

Logic gates do rather more than merely switch 'on' or 'off' for a particular input. They are designed to switch only when a certain set of conditions occurs. These conditions are defined by the Boolean functions 'AND', 'OR', and 'NOT', or combinations of these functions. These were considered in chapter 2, but for convenience, the definitions are repeated:

An AND gate gives an output only when signals at a specified level are present at *all* inputs.

An OR gate gives an output if there is a signal at a specified level at *one or more* inputs.

An inverter or NOT gate gives no output when an input is present and vice versa.

In all logic circuitry two voltage levels are used: *high* and *low,* the former being more positive than the latter. When describing the circuit function these levels are described as 1 and 0; if high is called 1 and low is called 0, the logic is said to be *positive logic*; if high is called 0 and low is called 1 the logic is *negative logic.* It is essential to state which kind of logic is being used if 1 and 0 describe the levels, because the gate function is different for the two kinds. (This is further clarified during the discussion of practical circuits.) A table describing the action of a logic gate is called a *truth table* and with each circuit examined the relevant truth table will be given.

Logic circuitry is classified by the type or arrangement of components used within it. Each type is called a *family*. Thus, there is the diode logic (DL) family, the diode-transistor logic (DTL) family and so on. Other families in popular use include transistor-transistor logic (TTL), resistor-transistor logic (RTL), emitter-coupled logic (ECL) and metal-oxide-semiconductor logic (MOSL). In addition, various trade names are used to describe logic families. Usually, a logic system uses only one kind of family; but, increasingly, one meets hybrid systems containing more than one as the manufacturers succeed in improving the ease of interconnection (interface). Important characteristics of logic families used to compare the different kinds are *fan out,* which is a measure of the number of gates that can be *driven* by one gate; *signal noise immunity* (SNI), which is a measure of the gate's sensitivity to random signals (noise); and *propagation delay,* which determines the speed of operation of the gate.

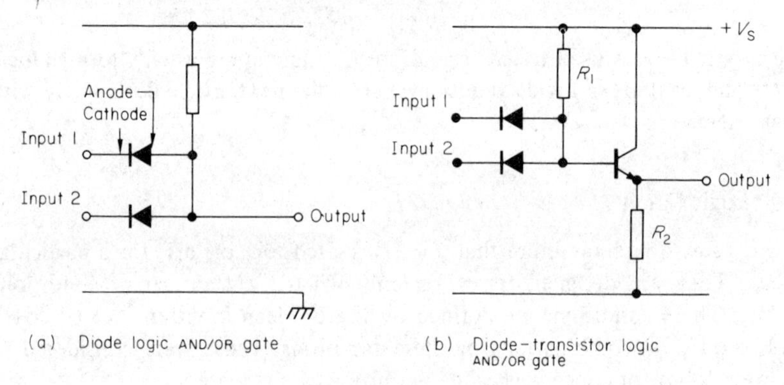

figure 9.4 DL and DTL gates

Figure 9.4a shows a simple two-input diode logic gate. The principle is that the voltage level at the anode of a diode that is conducting is the same as that at the cathode if the forward voltage drop is disregarded. Thus if a low level is applied to either input the diode will conduct and the output will be low; and only if a high level is applied at *all* inputs will the output be high. The gate truth table (table 9.1) is

Input 1	Input 2	Output
Low	Low	Low
High	Low	Low
Low	High	Low
High	High	High

table 9.1

Taking high as 1 and low as 0 (that is positive logic), the truth table (table 9.2) reads

Input 1	Input 2	Output
0	0	0
1	0	0
0	1	0
1	1	1

table 9.2

and the output is 1 only when all inputs are 1; that is, the gate is an AND gate. If negative logic (that is, high is 0 and low is 1) is used, the truth table (table 9.3) is

Input 1	Input 2	Output
1	1	1
0	1	1
1	0	1
0	0	0

table 9.3

and the output is 1 when one (or more) input is 1, so the gate is an OR gate. As the output is 1 when all inputs are 1, the OR function includes the AND function and the overall function is called the *inclusive* OR.

Diode logic is fairly slow and has a low fan out. It is however very simple to set up and maintain.

Figure 9.4b shows a diode-transistor logic gate. The transistor is forward biased via resistor $R1$ and the output is taken from the emitter, the emitter load being $R2$. If there is a high voltage level at each input, the transistor conducts and the output, being the voltage across $R2$, is high. All input diodes are reverse-biased under these conditons. Should a low voltage be applied to any one input, the respective diode conducts, robbing the transistor of its bias current. The transistor cuts off and the voltage at the output falls to zero (that is, a low level). The gate truth table (table 9.4) for two inputs is thus

Inputs		Output
1	2	
Low	Low	Low
Low	High	Low
High	Low	Low
High	High	High

table 9.4

Taking high as 1 and low as 0, it can be seen that the output is 1 only when all inputs are 1; that is, the gate is an AND gate. If negative logic is used (high as 0 and low as 1), the output is 1 whenever one or more inputs is 1; so the gate provides the OR function. If the load is transferred to the collector lead and the output is taken from the collector, the gate becomes a NAND (NOT AND) gate for positive logic and a NOR (NOT OR) gate for negative logic. It can be seen that the emitter follower connection provides inversion (that is, the NOT function) for what was a NAND gate, becomes an AND gate when the load is connected in the emitter. Because of its slow speed (high propagation delay) diode transistor logic is now less widely used, though it does have a fairly high SNI and a fairly large fan out.

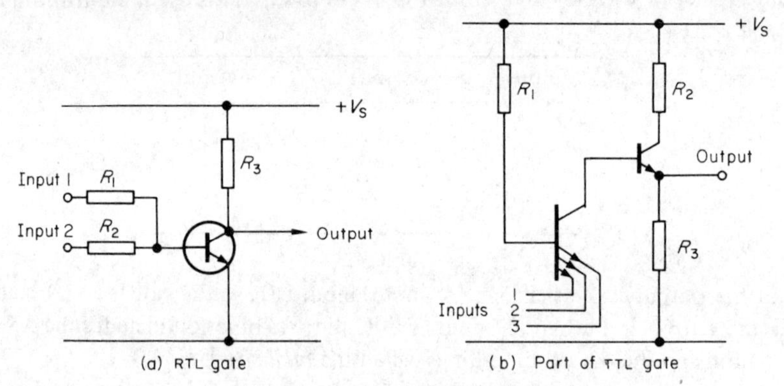

figure 9.5 RTL and TTL gates

Figure 9.5 shows RTL and TTL gates that provide either the NAND or NOR function depending upon the type of logic used. In figure 9.5a transistor bias is provided by one or more of the input resistors whenever the appropriate input is at a high level. If all inputs are low, the transistor is cut off and the output is high. The truth table (table 9.5) for two inputs is

Input 1	Input 2	Output
Low	Low	High
Low	High	Low
High	Low	Low
High	High	Low

table 9.5

For positive logic this is the NAND function and for negative logic the NOR function. Resistor-transistor logic has limited use because of its low speed, low SNI and small fan out.

Transistor-transistor logic is a relatively high speed logic family with a large

SNI and medium fan out. A TTL circuit is easily recognised by the multiple emitter transistor used at the input. Part of an integrated-circuit TTL NAND/NOR gate is shown in figure 9.5b.

Emitter coupled logic (ECL) is a very high-speed family with propagation delays of less than one nanosecond. It is characterised by a shunt connection of transistors as shown in the part-circuit illustrated in figure 9.6. The fan out is high, but the SNI is not as good as it is with the slower gates.

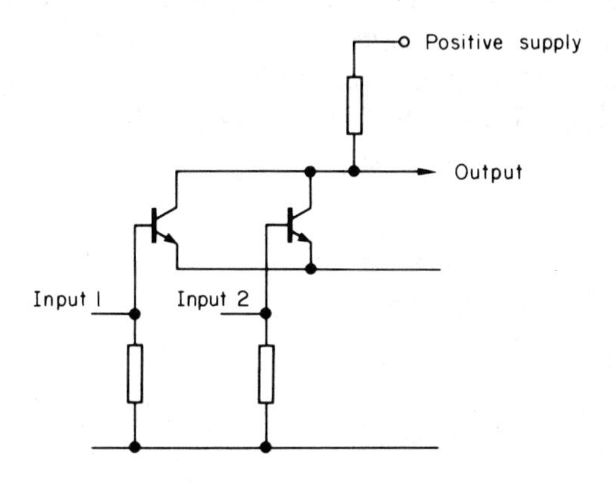

figure 9.6 Part of ECL gates

Logic families using field effect transistors (MOSL) are now available. Its chief advantage lies in the smallness of the integrated circuit form, which has a high density of gates per unit volume. At the time of writing MOSL is the slowest of the logic families.

9.5. Counting Methods

Slow speed counting using electrical circuits is achieved using an electro-mechanical counter not unlike the odometer ('mileometer') used in motor vehicles. In the electrical version pulses of current are applied to a solenoid which attracts an armature. The armature is linked to the unit counter which is thereby moved one unit per pulse. On the tenth pulse the unit counter wheel catches the tens counter wheel and this wheel moves one digit. Electro-mechanical counters are widely used in applications where speed of operation is not essential. Higher speeds may be achieved using electronic circuits and some form of glow-discharge tube to provide the read out, where this is necessary.

A circuit very commonly used as a basis for counting systems is the bistable multivibrator or 'flip-flop'. This is considered in the next article.

9.6. Bistable Multivibrator

A bistable multivibrator is made up of two valves or transistors acting as switches, the circuit as a whole having two stable states. In one state, one valve or transistor is fully conducting and the other is cut off; and, in the other state, the valve or transistor formerly conducting cuts off, and the valve or transistor formerly cut off conducts. Changing from one state to another is achieved by feeding a trigger pulse to the circuit for a time sufficient to start the switching process. Once started, the process continues by *cumulative action* until the other stable state is reached. Consider the circuit of figure 9.7a. In this circuit if $Tr1$ is fully conducting, the collector voltage is low and,

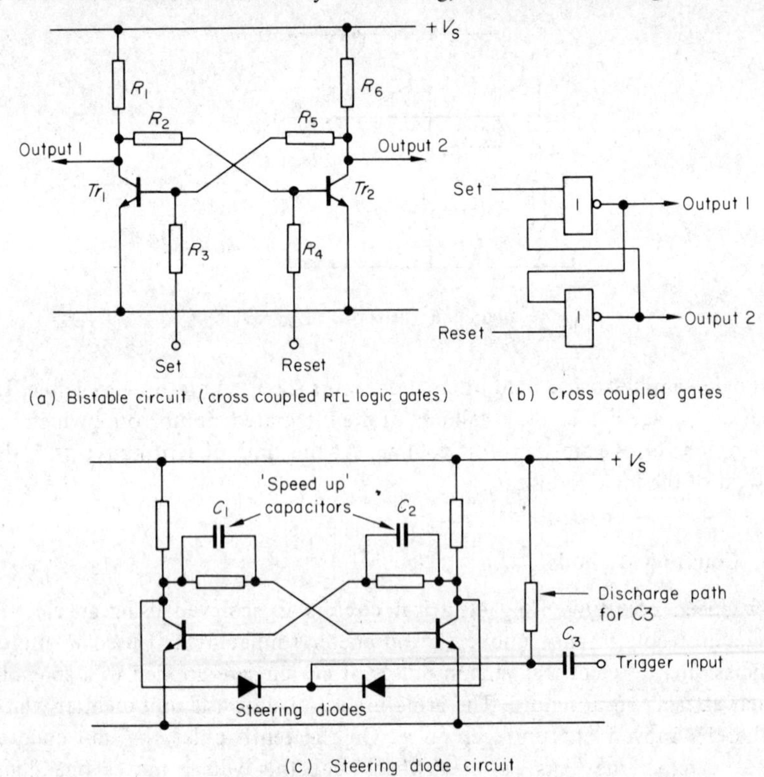

figure 9.7 The bistable multivibrator

because this voltage is fed to the base of $Tr2$ via $R2$, transistor $Tr2$ is cut off. The resultant high collector voltage of $Tr2$, fed to the base of $Tr1$ via $R5$, holds $Tr1$ fully conducting. This is one of the two stable states. If now a sufficiently large negative voltage is fed to the 'set' terminal (and thus to the base of $Tr1$ via $R3$) $Tr1$ begins to cut off. The rising collector voltage of $Tr1$, fed to the base of $Tr2$, causes $Tr2$ to begin conducting. The falling collector voltage of $Tr2$, fed to the base of $Tr1$, will continue the cutting off process of $Tr1$, even though the initial trigger pulse at the set input may have gone. The switching process continues until $Tr1$ is completely cut off and $Tr2$ fully conducting. The circuit now remains in this state until a trigger pulse is applied to the 'reset' terminal. The circuit shown in figure 9.7b portrays the unit as two cross-coupled RTL NOR gates. Bistable multivibrators may be similarly made up using two gates of any of the other logic families. The circuit shown is called an RS or SR *flip-flop,* R standing for *reset* and S for *set.* The inclusion of a *steering diode* circuit as shown in figure 9.7c replaces the two trigger input lines by a single line, the diodes guiding or *steering* the input pulse to the non-conducting transistor as determined by which stable state the flip-flop is in. The negative trigger pulse causes the diode between the input and the high collector (the off transistor) to conduct, driving the collector voltage downwards and starting the switching process. Such a circuit is called a T (trigger) flip-flop. The steering diode circuit may alternatively be attached to the transistor bases instead of the collectors. Capacitors $C1$ and

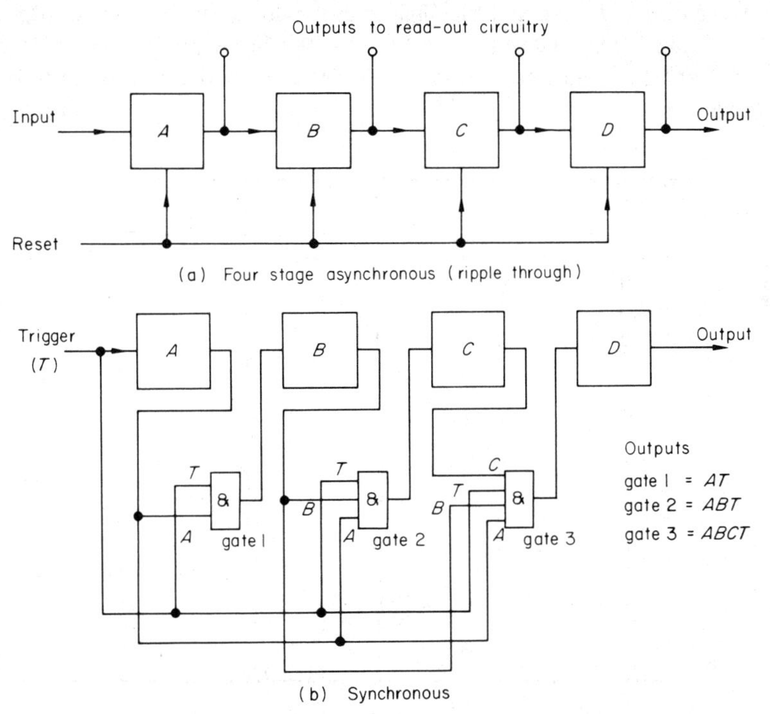

figure 9.8 Scale-of-16 binary counters

*C*2 in the circuit shown reduce the switching time of the circuit so allowing higher speeds to be used.

9.7. Counter Circuits

A four-stage pure binary counter is illustrated in figure 9.8a. It consists of four flip-flops (shown in block form) cascaded, so that each flip-flop provides a drive pulse to the next and receives its own drive pulse from the previous circuit. The circuit is arranged so that each flip-flop switches when it receives a negative going pulse (that is, when the previous output goes from a high to a low voltage level). Consider the circuit when all outputs are low. The high-to-low edge of the first input pulse causes *A* to switch and its output goes high, though this has no effect on flip-flop *B* because the direction of the pulse into *B* is low-to-high. The second input pulse again switches *A*, the *A* output goes low and therefore switches *B* whose output then goes high. All outputs except *B* are now low. The third input pulse switches the *A* output from low to high (again having no effect on *B*) so that *A* and *B* outputs are now high. The fourth input pulse switches output *A* low, which in turn switches output *B* low, and this high-to-low pulse from *B* switches output *C* high. If 'high' is denoted by 1 and 'low' by '0' (that is, positive logic), a truth table (table 9.6) showing all the possible combinations of outputs is as follows.

D	*C*	*B*	*A*	Input pulse
0	0	0	0	0
0	0	0	1	1
0	0	1	0	2
0	0	1	1	3
0	1	0	0	4
0	1	0	1	5
0	1	1	0	6
0	1	1	1	7
1	0	0	0	8
1	0	0	1	9
1	0	1	0	10
1	0	1	1	11
1	1	0	0	12
1	1	0	1	13
1	1	1	0	14
1	1	1	1	15

table 9.6

As can be seen, after the 15th input pulse all outputs are high and the next input pulse will reset all flip-flop outputs to low, producing a high-to-low pulse at output *D*. The counter as a whole thus produces one output pulse for every sixteen input pulses. It is therefore described as a 'scale of sixteen' binary counter. The word binary means having two states or two digits. The truth table (table 9.6) shown gives the *binary equivalent* of decimal numbers 0 to 15; decimal 7 for example, being written as 111, decimal 8 as 1000. Examination of the table shows that flip-flop *A* switches in response to each input pulse, flip-flop *B* in response to every 2 input pulses, flip-flop *C* in response to every 4 input pulses and flip-flop *D* in response to every 8 input pulses. The counter shown is said to use an 8421 *code* for this reason. The binary number 0101 thus means

$$(0 \times 8) + (1 \times 4) + (0 \times 2) + (1 \times 1)$$

which is 5; that is, five input pulses have been received by an 8421 counter of the type shown. The binary numbering system is widely used in computing and other pulse circuits because only two digits are involved and these can be represented by easily distinguishable electrical states 'on' and 'off' (fully conducting and non-conducting). Use of the decimal system (ten digits, 0 to 9 inclusive) would require ten distinguishable levels, a rather more difficult state of affairs. The simple counter shown will have a delay in counting because each switching process takes time. Flip-flop *D* for instance cannot change until *A*, *B* and *C* have all switched. The switching pulse is said to *ripple through* the system and the counter is called a *ripple-through* or *asynchronous* counter. The speed of operation can be considerably improved by synchronising all flip-flops to switch simultaneously. This produces a *synchronous* counter as shown in figure 9.8b.

Examination of the truth table (table 9.6) shows that when output *A* is 1

D	*C*	*B*	*A*	Input pulse
0	0	0	0	0
0	0	0	1	1
0	0	1	0	2
0	0	1	1	3
0	1	0	0	4
0	1	0	1	5
0	1	1	0	6
0	1	1	1	7
1	0	0	0	8
1	0	0	1	9

table 9.7

and a trigger pulse is received, output B changes. When output A and output B are 1 and a trigger pulse is received, output C changes. When outputs A and B and C are all 1 and a trigger pulse is received, output D changes. The arrangement illustrated in figure 9.8b satisfies all these conditions. Consider flip-flop C, for example; if outputs A and B are 1 the AND gate feeding input C has a 1 at all of its 3 inputs. Immediately a trigger pulse (that is, a change from 1 to 0) is received the output of the AND gate changes from 1 to 0 and C switches *without* having to wait for the change to ripple through the previous flip-flops. Thus the count is synchronised with the trigger pulse (disregarding the propagation delay of the AND gates).

The counters shown count in pure binary, four cascaded flip-flops giving an output after 16 (2^4) pulses have been received (3 flip-flops give an output after 2^3 pulses, 2 flip-flops after 2^2 pulses). The counter is called a 'scale of 16' counter, as stated above, for that reason. A common requirement, since our number system is decimal, is for a counter to count in tens. Batch counters may be required to count batches of any number of items ranging from 2 upwards; a particular example (at least prior to metrication) being the requirement to count in dozens. Similarly, if a counter system is being used in

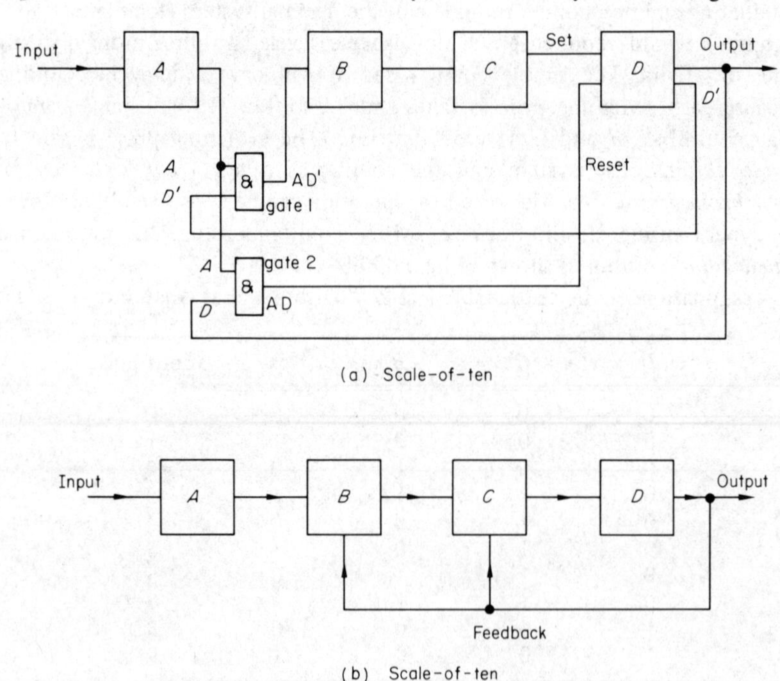

(a) Scale-of-ten

(b) Scale-of-ten

figure 9.9 **Binary coded decimal counters**

frequency division (that is, to reduce the frequency of a given input in a particular ratio) the ratio required may not be a function of 2. One example of this is the digital clock that is mains driven at 50 Hz but requires pulses every 60 seconds, and every 60 minutes, and so on. For such purposes, *binary coded decimal* or *BCD* counting systems are employed.

Figure 9.9a shows a *BCD* (scale of ten) ripple-through counter. Flip-flop *B* is fed via AND gate 1 which has two inputs, one from output *A*, the other from output D' (that is, the output of flip-flop *D* that is low when the normal output line *D* is high). The reset input of flip-flop *D* is fed via AND gate 2 which also has two inputs; one from output *A*, the other from output *D*. Consider now the truth table (table 9.7) of an 8421 counter for the first 9 pulses.

Up to and including pulse 7, the D' line is 1 (the complement of *D*) so gate 1 allows normal switching to proceed. Line *D* is 0 so the reset line from gate 2 cannot be activated. When the 8th pulse has been received and has rippled through to change *D* to 1, D' changes to 0. The output of AND gate 1 then changes to 0, and thus any subsequent 1 to 0 pulses will not be transmitted to *B* or *C*, which will therefore remain at 0. AND gate 2 now has inputs of 1 on the *D* line and 0 on the *A* line, the output of AND gate 2 is thus 0. At the ninth pulse *A* changes to 1, AND gate 2 opens and its output also goes to 1. At the tenth pulse *A* changes from 1 to 0, so the output of AND gate 2 changes from 1 to 0, thereby providing a trigger pulse for the reset of flip-flop *D*. Output *D* now goes from 1 to 0 (giving an output pulse in the process) and all outputs are now 0 ready for the decimal count to start again.

Figure 9.9b shows another form of *BCD* counter using a 2421 code and using feedback to reset outputs to 0 at the tenth pulse.

The truth table (table 9.8) for this circuit is

D	*C*	*B*	*A*	Pulse
0	0	0	0	0
0	0	0	1	1
0	0	1	0	2
0	0	1	1	3
0	1	0	0	4
0	1	0	1	5
0	1	1	0	6
0	1	1	1	7
1	1	1	0	8
1	1	1	1	9

table 9.8

The decimal equivalents of the flip-flop outputs (this will correspond to the number of input pulses received) are

$$D = 2 \qquad C = 4 \qquad B = 2 \qquad A = 1$$

Thus 9 which is 1111 is

$$(1 \times 2) + (1 \times 4) + (1 \times 2) + (1 \times 1)$$

(In the 8421 (pure binary) code 9 is written 1001 which is

$$(1 \times 8) + (0 \times 4) + (0 \times 2) + (1 \times 1))$$

To obtain this truth table feedback lines are taken from output D to flip-flops B and C. On receipt of incoming pulses the count proceeds normally until the 7th pulse has been received. The outputs $DCBA$ are now 0111 (that is,

$$(0 \times 2) + (1 \times 4) + (1 \times 2) + (1 \times 1)).$$

At the 8th pulse A changes from 1 to 0 which changes B then C then D. Without feedback the binary number (8421 code) would now be 1000. However, the change in D is fed back to B and C, at such a point that they both return to 1. This gives 1110 for the 8th pulse (corresponding to pulse 14 in an 8421 counter). The 9th pulse changes A to 1 and the counter now reads 1111 (a state reached after 15 pulses in the 8421 counter). The 10th pulse resets all outputs to 0, giving a pulse (as D changes from 1 to 0) at output D. The count is made to 'jump' by 6 by feedback to flip-flops B and C (that is, $2^2 + 2^1$ added pulses).

Many other counter circuits are available (in a variety of *BCD* codes) to meet specific requirements. Among these are two particular forms of counter known as ring counters and shift registers. A *ring counter* comprises a standard synchronous counter in which the output bistable is connected back to the input bistable to permit continuous circulation of pulses to the full capacity of the counter. For example, a scale-of-ten counter would continuously re-circulate 10 pulses, and at each complete pass the output pulse could be used to trigger a second scale-of-ten ring counter, thus forming a scale of 100. Such a device forms an electronic equivalent to the mechanical device mentioned in section 9.5. A *shift register* is a collection of bistables used to hold a binary number, and which can be pulsed in such a way that the entire binary number moves one stage to the left or one stage to the right.

10 Transmission and propagation

This chapter is concerned with the transmission and propagation of electronic signals using conductors and using electromagnetic radiation. In telecommunication systems conductive lines are used to carry signals between systems and from aerial to system or vice versa. Electromagnetic radiation is used between transmitter and receiver, when these are separated by distances that do not allow the use of conductive lines. Particular examples include radio and television systems and telemetry systems used with missiles and space vehicles.

10.1. Transmission Lines

Transmission lines are conductive paths by which electrical energy is transferred from one point to another. For effective transfer the energy loss in the line must be as small as possible and energy should be transferred in one direction only. There should therefore be no reflection of energy back along the line. Conductors have internal resistance, capacitance and inductance at all frequencies but the reactive effects are particularly important at high frequencies. Any transmission line can be considered to be made up of a number of inductive-capacitor sections as shown in figure 10.1. Such a representation is said to be a line with *lumped constants.*

An important characteristic of any transmission line is the ratio of voltage to current at the line input, and also at points along the line. The former impedance is called the input impedance; the latter is called the *characteristic impedance* and is symbolised Z_0. For an infinitely long line the input impedance is equal to the characteristic impedance and its value is given by $\sqrt{(L/C)}$, where L and C are the values of inductance and capacitance, respectively, per unit length of line. There is no reflection of energy along such an infinite line and, because power is being continuously absorbed by the line, it can be regarded as being resistive (having zero phase shift between line-current and line-voltage).

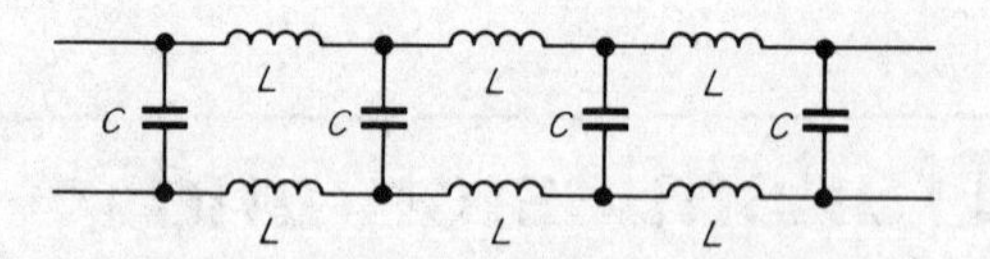

figure 10.1 Transmission line equivalent circuit (lumped constants)

In practice transmission lines are of finite length but if steps can be taken to *simulate* the performance of an infinite line, energy transfer is considerably improved. These steps, which involve selecting a certain length of line for a particular signal frequency and terminating the line in an impedance of a certain value, will now be considered.

If an infinitely long transmission line is broken (as in figure 10.2) to give a short finite line disconnected from the remainder of the infinite line, the

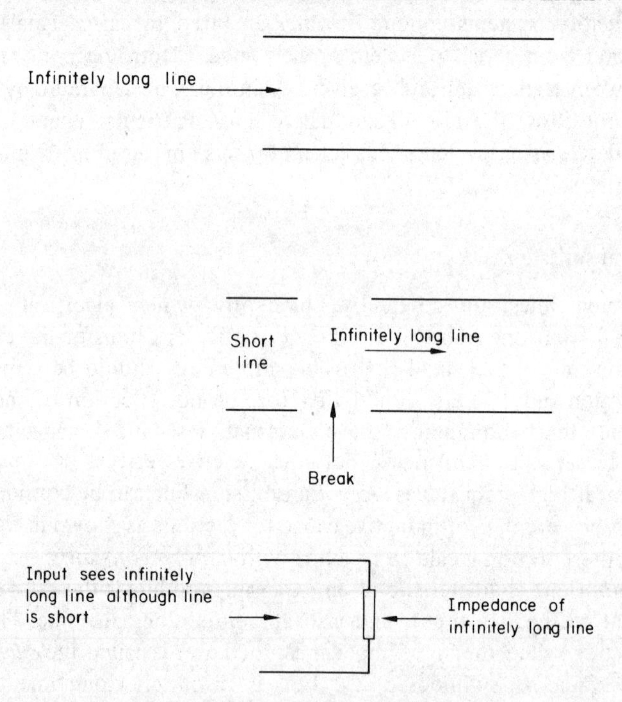

figure 10.2 Infinite transmission lines

input conditions to the residue of the infinite line at the break point are still those of an infinite line. Thus, if this residue of the infinite line is replaced by a pure resistance of value $\sqrt{(L/C)}$ (that is, the input impedance of the infinite line) the finite line is terminated in what is apparently an infinite line. Thus infinite line conditions of energy transfer are simulated in the finite line. The finite line is now correctly terminated and there will be no reflections back along the line.

If a finite line is terminated in any other value of impedance, voltage waves and current waves are reflected back from the receiver and these react with the forward or *incident* voltage and current waves to give *standing* voltage and current waves along the line. The positions along the line at which maximum or minimum values of the standing waves occur at any given instant do not alter, although the actual values may do so depending on the relative positions of the incident and reflected waves, which (respectively) move down and up the line. A suitable meter to measure the standing wave would indicate a different value at different points along the line, giving a maximum where standing wave peaks occur and a minimum where standing wave troughs occur. In a correctly terminated line there is no reflected wave and thus only the incident wave travelling down the line. A meter would thus indicate a constant value at all points along the line, the value being the r.m.s. value of the incident (signal) wave.

The *standing wave ratio* (SWR) is the ratio of the maximum value of the standing wave to the minimum value. The most severe case of a mismatched line occurs when the line is terminated in a short-circuit or in an open-circuit. In these extreme cases the standing wave maximum equals twice the maximum value of the incident wave and the minimum value is zero. Thus the SWR is infinity. For a correctly terminated line there is no standing wave and, as stated above, a meter reads the same value all along the line. Thus the maximum and minimum readings are the same and the SWR is unity. This is the ideal value of the standing wave ratio. (See figure 10.3.)

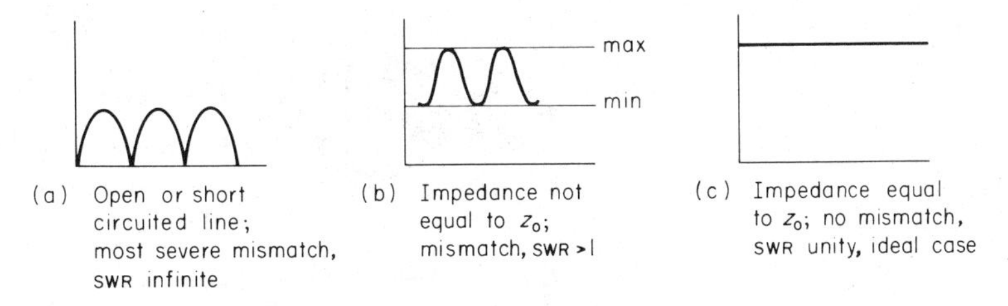

figure 10.3 Standing waves on transmission lines

The *velocity of propagation* of the incident wave down a transmission line is the speed at which the wave travels down the line. It is determined by the inductance and capacitance per unit length of the line; and, if L and C denote these properties, the velocity of propagation is given by $1/\sqrt{(LC)}$ for r.f. transmission. The *propagation coefficient* of a transmission line is the ratio between received voltage and transmitted voltage. Because of the signal attenuation and phase shift that occurs in the line the propagation coefficient is made up of two parts, the *attenuation coefficient* and the *phase change coefficient.* For maximum efficiency of transmission the attenuation should be as *low* as possible.

Three types of transmission line in common use are the open twin feeder, the single coaxial cable, and the shielded pair feeder. The open twin feeder consists of a pair of conductor wires separated by low loss dielectric spacers as shown in figure 10.4a. This type of line has a relatively high characteristic impedance and can handle large transmitted power levels. The coaxial cable shown in figure 10.4b consists of a single conductor surrounded by dielectric upon which the second conductor is wrapped in braided copper form. The overall cable is plastic covered. The characteristic impedance of the cable is commonly about 75 ohms, but other values are possible in the range 30 to 100 ohms. The advantages of this type of line include low radiation loss and

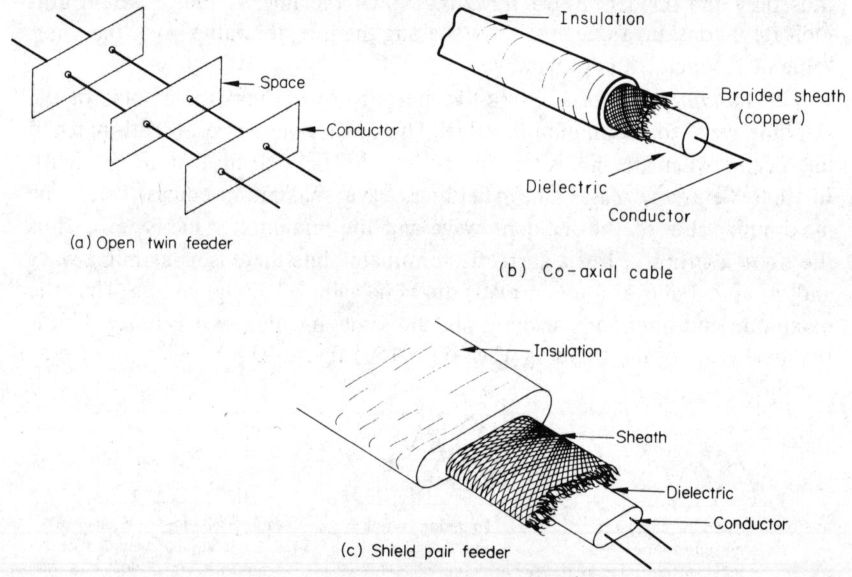

figure 10.4 Types of transmission line

extreme flexibility in use. One disadvantage is the fact that the line is *unbalanced* because the outer conductor is normally earthed to provide a screen for the inner conductor. Unbalanced lines can introduce matching problems. Figure 10.4c shows a shielded pair feeder, similar in construction to single coaxial cable. The characteristic impedance is similar to that of the single coaxial cable and the line is balanced. However, such a construction reduces the power handling capabilities of the line.

For normal energy transmission it is necessary to terminate the line as correctly as possible to avoid mismatch, reflected waves, and thus standing waves. A mismatched line returns some energy back down the line with consequent loss. In some cases, however, short lengths of deliberately mismatched line are used because of their variable impedance properties. Figure 10.5 shows the standing voltage and current waves along a line equal in length to one wavelength of the transmitted signal. (The wavelength depends on frequency and is given by velocity of electromagnetic radiation divided by frequency). Part (a) shows standing waves on an open circuited line and part (b) shows standing waves on a short circuited line.

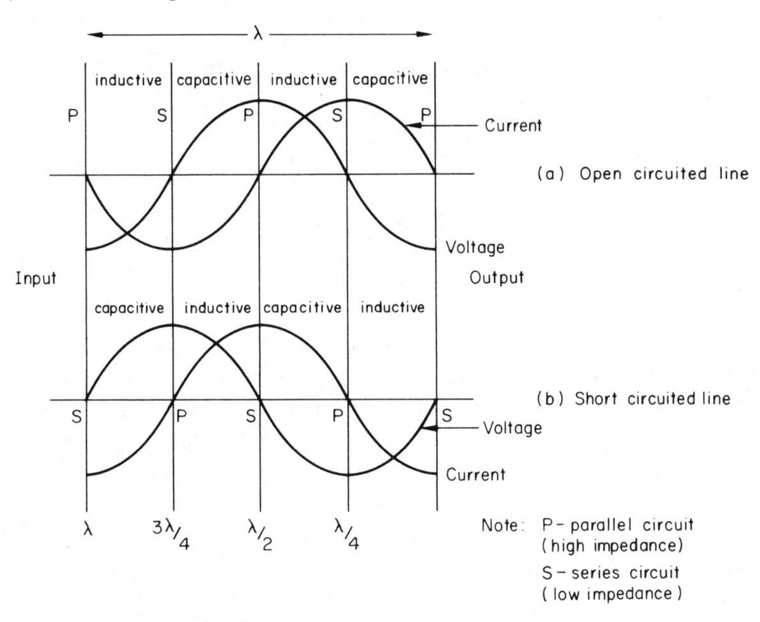

figure 10.5 Behaviour of short lengths of line

Examine figure 10.5a. The output shows maximum voltage and zero current which is consistent with an open circuit. The input end shows a similar state of affairs and the impedance at this point is a *maximum*. In this respect the line *at this point* appears as (that is, has the same effect as) a parallel tuned circuit. This set of conditions occurs again at $\lambda/2$ and thus an open circuited line of length λ or $\lambda/2$ behaves as a parallel tuned circuit at its input. At the $3\lambda/4$ and $\lambda/4$ points the opposite state of affairs occurs with zero voltage and maximum current; this gives minimum impedance similar to a series tuned circuit at resonance. Thus open circuited *three-quarter wave* or *quarter wave* lines behave as series tuned circuits. In between these points the line presents either an inductive or capacitive reactance, determined by the sign of the ratio between voltage and current. Between λ and $3\lambda/4$, for example, both voltage and current are shown negative which indicates a positive inductive reactance. Between $3\lambda/4$ and $\lambda/2$, voltage is positive, current negative and the effective impedance is negative (that is, capacitive). Examination of figure 10.5b for a short-circuited line gives similar results but with the inductive/capacitive effects and series-tuned-circuit/parallel-tuned-circuit effects reversed. A summary of behaviour (table 10.1) is as follows.

Line length	Open circuit termination	Short circuit termination
Less than $\lambda/4$	Capacitive	Inductive
$\lambda/4$	Series circuit	Parallel circuit
$\lambda/4$ to $\lambda/2$	Inductive	Capacitive
$\lambda/2$	Parallel circuit	Series circuit
$\lambda/2$ to $3\lambda/4$	Capacitive	Inductive
$3\lambda/4$	Series circuit	Parallel circuit
$3\lambda/4$ to λ	Inductive	Capacitive
λ	Parallel circuit	Series circuit

table 10.1

Thus, any type of reactance may be obtained by choosing an appropriate line length and suitable termination. Some of the uses of short lines are metallic insulators, matching transformers, matching stubs, and lecher bars.

(Note: It is essential to realise that the wavelength depends on frequency and thus what is a quarter wave line to one signal is not a quarter wave line to another.)

As shown above a short circuited quarter wave line or *quarter wave stub* presents a very high impedance at the input. Thus quarter wave stubs can be used to support the main transmission lines without seriously affecting the signal transmission since as far as signal is concerned the lines are not joined. This application appears unusual at first because the lines are apparently short-circuited. Quarter wave stubs used for line support are called *metallic insulators.*

The separation between conductors of a short length of line affects the capacitance per unit length and thus the input impedance. Insertion of a short line with adjustable separation between conductors between, say, the main transmission line and aerial of a radio system effectively provides a variable impedance between the two parts of the system. Best matching between line and aerial can thus be achieved by adjusting the impedance of the inserted variable-separation line. The inserted length which is usually quarter wave behaves much like a *matching transformer* used at lower signal frequencies.

Maximum power dissipation at the end of a transmission line occurs if the load is purely resistive. Mismatched lines set up standing waves that produce impedances (consisting of resistance and reactance) at various points along the line. By attaching short lines (called *matching stubs*) having the opposite kind of reactance at these points the impedance can be reduced to one containing pure resistance. Thus maximum transfer conditions can be set up along the line.

Lecher bars are short lengths of shorted-line that are used to take the place of conventional tuned circuits at high frequencies. As was noted in chapter 7, reactive components at high frequencies become too small for convenient handling and the capacitance and inductance of a transmission line can be used instead of discrete components. Oscillators at the high frequency end of the spectrum invariably use lecher bars or an equivalent device in place of conventional tuned circuits.

10.2. Propagation by Electromagnetic Waves

Whenever electric current flows in a conductor an electric field and a magnetic field are present in the vicinity of the conductor. If the current is alternating in nature both fields will also alternate, their strengths rising and falling with time. The combination of these fields, which act at right angles to one another (as shown in figure 10.6), produce an electromagnetic disturbance

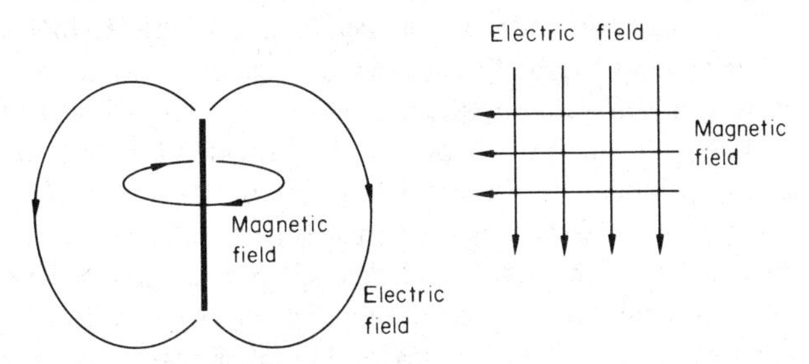

figure 10.6 Electromagnetic field patterns for propagation

or wave that is capable of energy transference from one point to another. The effectiveness of the transference depends upon a number of factors, particularly frequency; but it should be noted that electromagnetic radiation occurs at all frequencies, even as low as that of the mains electricity supply. The effectiveness of the radiation depends also on the type and length of *aerial* used (as described in article 10.3); the higher the frequency, the shorter the aerial for best transfer. Mains cables often provide an effective aerial leading usually to undesirable mains radiation. As shorter aerials are obviously more convenient to handle, the higher end of the electromagnetic spectrum is normally used for radio transmission. Other factors taken into consideration in choosing a particular carrier frequency include the method of propagation, which varies with the frequency of the transmitted wave, as described below.

Electromagnetic waves are propagated in two ways, one of which can be further divided into two. These are the *ground wave* and the *sky wave*; the ground wave consists of two further types, the surface wave and the space wave.

The surface wave travels relatively close to the earth's surface following the curvature of the earth. The space wave follows the direct line between transmitter and receiver so that space wave transmission is referred to as 'line of sight'. If the receiver is below the horizon as far as the transmitter is concerned transmission by space wave alone is not possible. The surface wave component of the ground wave then has more effect.

The sky wave is radiated directly into the upper atmosphere and, depending upon the *angle of incidence* and the wave frequency, may be reflected back to earth by one or other of the various layers of charged particles that exist above the earth, and are collectively called the *ionosphere.* The angle of incidence is the angle between the line of propagation of the wave and a line drawn vertically to the surface of the charged layer. The ionosphere itself, which is composed of several layers termed *D, E* and *F,* is set up by atmospheric particles that gain or lose charge due to the sun's radiation. The height of the various layers above the earth's surface is determined particularly by the time of day or night and also the time of year, because this in turn is affected by the position of the sun relative to the earth.

Whether or not the sky wave or part of it is reflected is determined by two factors; the angle of incidence (as previously described), and the frequency of the propagated signal. If the angle of incidence is reduced, there comes a point at which the incident rays travel through the ionosphere layer and are lost. This angle of incidence is called the *critical angle.* The reflective or, to be precise, refractive (bending) properties of the layer (as regards a particular wave) depend to some extent on frequency; and, if the frequency is progressively increased, eventually the incident wave is not reflected but is instead lost in the charged layer. The frequency at which this occurs is called the *critical frequency.*

At frequencies between 10 kHz and 300 kHz (wavelength 30 km to 1 km), called the *long wave band* or *Band A,* transmission is mainly by ground wave, particularly the surface wave. Attenuation of the surface wave (due to absorbed energy caused by induced surface currents) is least severe over surfaces having large conductivity. Consequently, this band is especially useful for maritime communication. Transmission at these frequencies is however subject to interference from atmospheric disturbances and from unsuppressed electrical equipment.

Between 300 kHz and 3 MHz (wavelength 1000 m to 100 m) transmission over short ranges is mainly by ground wave, and over longer ranges (up to 1000 miles) by sky wave. This band is called *Band B* or the *medium wave band.*

In *Band C,* the *short wave band,* which lies between 3 MHz and 30 MHz, transmission is mainly by sky wave, which is reflected by the ionosphere. The distance covered by the sky wave from transmitter to receiver when the maximum usable frequency is being employed is called the *skip distance.* Reception is possible a short distance from the transmitter by means of the ground wave, the distance between the final point of ground wave reception and the point of sky wave reception being called the *dead space.* See figure 10.7. This band is not as susceptible to interference as bands *A* and *B.*

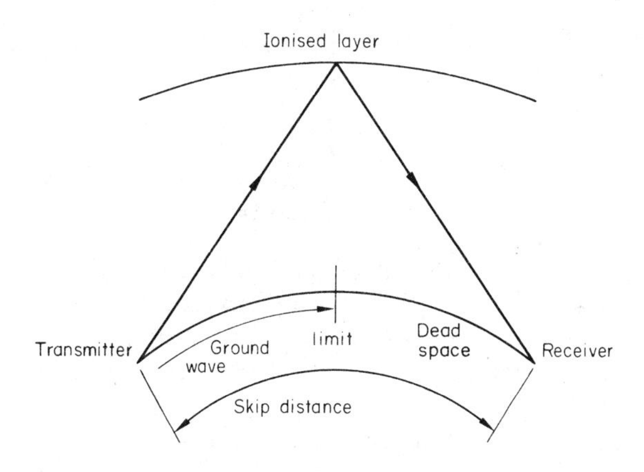

figure 10.7 Skip distance etc.

Band D, 30 MHz to 300 MHz, uses space wave propagation, because the ground wave suffers marked attenuation and the sky wave is not returned in many cases. This band is used for short range television and VHF broadcasting, the receiver and transmitter preferably not being separated by terrain changes (hills etc.) or other obstacles if good transmission is to be obtained.

Band E which lies above 300 MHz is used in radar systems and, as with Band *D,* line of sight transmission using the space wave is employed.

10.3. Aerials

A system to transmit or receive electromagnetically propagated signals is only as efficient as its aerial. Correct matching of transmission lines between aerial and system and the use of high gain amplifiers providing maximum range and quality are no use alone unless the correct aerial is used with the system.

Resonant aerials are so called because they behave in a similar manner to a series resonant or acceptor circuit. Two of the main kinds of resonant aerial are the half-wave dipole and the Marconi quarter-wave aerial illustrated in figure 10.8. The half-wave aerial, which was originally developed by Hertz in

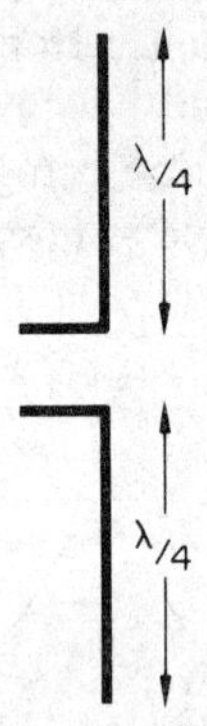

figure 10.8 Marconi quarter-wave aerial

1887, has a length equal to half of the wavelength of the radiated signal frequency, because it is found that with this length maximum resonance effect, and thus maximum effectiveness of radiation, is achieved. The quarter-wave aerial, developed from the Hertz design, uses the earth as a conducting plane.

Before considering aerial designs further, it is necessary to define certain terms used to describe their performance. These include *aerial resistance, matching, polarisation* and *polar diagram.*

Aerial resistance is the effective resistance of the aerial used in determining the energy dissipation in the aerial. The aerial resistance is not necessarily the same as the *aerial impedance,* which includes the inductive and capacitive reactance of the aerial and is determined by the point of connection to the aerial. A centre-connected half-wave dipole, for example, has a resistance and impedance of about 75 ohms; however if the dipole is end-connected, the resistance remains the same but the impedance may be of the order of 2500 ohms.

As was explained in the section on transmission lines, for maximum transfer of energy between aerial and line the line resistance and aerial resistance should be equal. The process of obtaining maximum transfer by arranging this state of affairs is called *matching.*

As also stated, an electromagnetic wave is set up by a magnetic field and an electric field acting at right angles to each other. A *horizontally polarised* aerial is one that has the associated electric field lying in the horizontal plane; a *vertically polarised* aerial has the electric field in the vertical plane. Figure 10.6 showed that the electric field plane lies along the aerial, so that a vertically mounted aerial is vertically polarised as shown in figure 10.9a. Figure 10.9b shows horizontal polarisation.

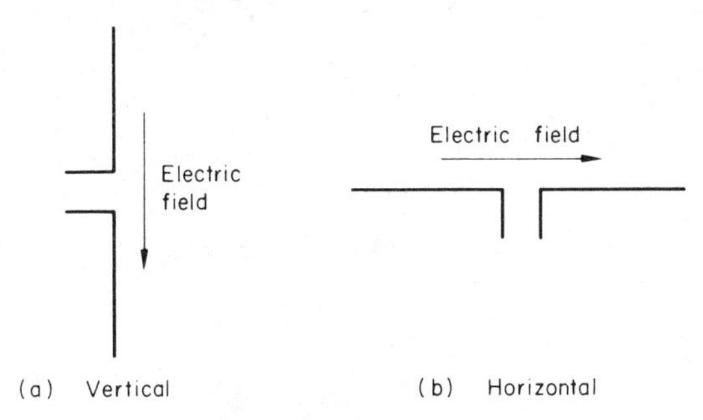

figure 10.9 Polarisation

If the electric field strength is measured at various points away from the aerial it is found that uniform radiation does not occur in all directions in both the vertical and horizontal planes. If lines are drawn on a diagram joining together all points where the observed field strength is the same, the resultant picture is called a *polar diagram.* A polar diagram for a transmitting aerial shows the direction in which the transmitted energy will be most strongly radiated; for a receiving aerial the polar diagram shows in which direction the aerial should be mounted to obtain best reception. Diagrams for a half-wave dipole are shown in figure 10.10. As can be seen, for a vertically mounted aerial, figure 10.10a, the vertical polar diagram shows zero radiation above and below the aerial and maximum radiation along the centre line of the aerial in the vertical plane. In the horizontal plane the field strength is uniform in all directions for the vertically mounted aerial, as shown by the circular polar diagram.

For a horizontally mounted aerial the diagrams are as shown in figure 10.10b and c.

The simple aerials shown are relatively non-directional, for radiation takes place equally in all directions in one plane or the other, depending on the mounting of the aerial. Introduction of additional elements called *parasitic elements* to the simple aerial produce a directional array such as the one

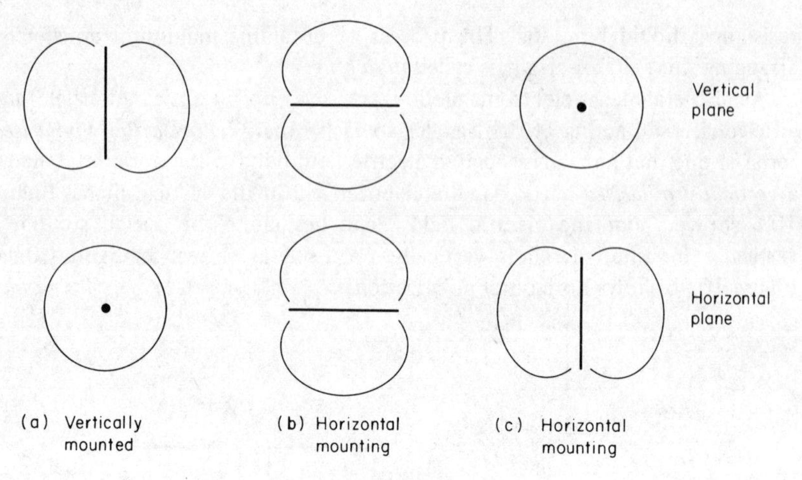

figure 10.10 Polar diagrams

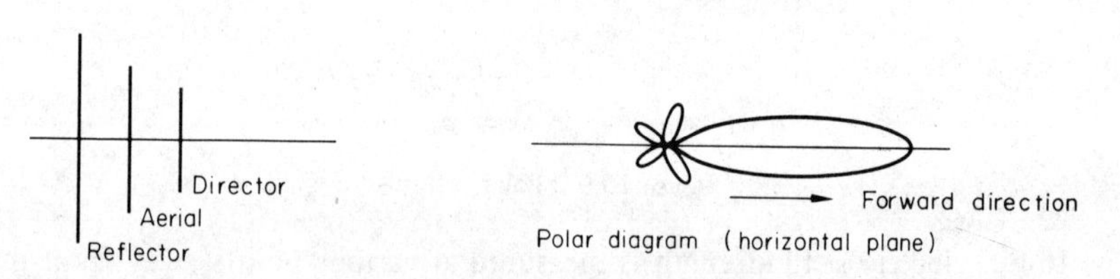

figure 10.11 3-element Yagi array

shown in figure 10.11. This aerial is a 3 element Yagi aerial with one reflector and one director as shown. The director is in front of the aerial and is shorter than the aerial itself; and the reflector, which is slightly longer, is situated *behind* the aerial. The resultant polar diagram in the horizontal plane shown in the same figure shows a highly directional aerial as opposed to the simple circle diagram of the aerial alone. Other Yagi arrays are possible with differing numbers of directors etc.

For maximum reception of any given signal the factors to be considered are, the frequency (and thus wavelength), the direction of polarisation, and (in certain cases, involving 'line of sight' transmission), the general direction in which the transmitter is situated. The appropriate aerial type, mounting and direction can then be selected.

11 System faults

A fault in a system is a change in the operating characteristic of one or more components which culminates in system failure. Failure in this context is defined as the inability of the system to carry out the function described by the system specification.

There are many different types of failure including sudden, gradual, partial, complete and so on. The meaning of these terms is fairly clear from the normal usage of the words. Two other terms commonly applied need defining, however; these are *chance failures* and *wearout failures.* A chance failure is one that occurs suddenly and at random and may occur after a short time in service or a long time in service. Wearout failure, on the other hand, is failure due to the wearing out of a component and would normally occur after a long time in service, as determined by the expected useful life of that component.

Three factors are taken into consideration when selecting components, either for the first time or as replacement items for those previously failed. These factors are *reliability, mean time between failures,* and *mean wearout life.* Figures for these quantities are usually given by the manufacturer. The *reliability* of a component is the probability that it will function properly over a particular period of time. Reliability is expressed as a fraction or as a decimal figure, the maximum value being unity. If the reliability were unity the implication would be that its chances of survival were completely certain. The *mean time between failures* (MTBF), as the name suggests, is the average time which elapses between chance failures. Being an average figure, one cannot expect it to be the exact time between failures, because by definition 'chance' means occurring randomly. Nevertheless, the figure can often be a useful guide to reliability. The *mean* (*wearout*) *life* is, of course, the expected useful life of a component from start of service to wearout (not chance) failure. As stated, these figures should be borne in mind if available when selecting components for system subunits.

11.1. Causes of Failure

Failure of a system when caused by an internal fault is invariably due to failure of one or more components within the system subunits. Component failure, in turn, is due to excessive stress acting on the component. Component stresses are of two kinds, environmental and operating; the first kind being stresses due to the surroundings in which the component is operating; the second kind being due to the circuit conditions under which the component is operating. Environmental stresses may be caused by atmospheric pressure (or the lack of it), humidity of the surroundings, chemical content of the surroundings, nuclear radiation (if present) and temperature; temperature is also affected by the operation of the system and thus in a sense is also an operating stress. Operating stresses are set up by circuit conditions and thus are determined by component voltage, current (and thus power) and frequency of operation. Stresses which are neither wholly environmental nor operating include those set up mechanically. These include vibration, friction, shock due to high acceleration, and so on.

Atmospheric pressure, which at ground level is usually taken as that indicated by a column of mercury 760 mm in height, varies considerably as the height above ground is increased and is virtually zero at levels above the stratosphere. Components in cans or glass containers are normally subject to the pressure within the container; if the pressure difference between the inner and outer regions of the container is such that container damage results, the component must then function at a new pressure. In certain cases this may result in complete failure.

Humidity is a measure of the content of water vapour in the air or surrounding gases. Since impure water is a relatively good conductor of electricity, components may suffer resistance changes which alter their operating characteristics. In the most severe cases this could lead to a short circuit. Atmospheric humidity in natural conditions may vary from 1 or 2% to 100% depending on geographical location. In addition, for industrial control systems concerned with certain factory processes, a humidity which is higher than usual may have to be tolerated and this factor must be borne in mind.

The chemical content of the surrounding air can have a considerable effect on normal functioning of components. Normal atmosphere is a nitrogen-oxygen mixture containing a very small percentage of hydrogen, but the relative proportions of the constituent gases varies considerably with height above ground level. The air may also contain heavy proportions of impurities including: sulphur dioxide, which combines with water vapour to produce sulphuric acid; soot, which is basically a conductive carbon compound; and 'dust', which may be made up of a variety of chemical compounds depending on location. Here again industrial control systems are often situated close to the heart of the manufacturing process and care must be taken to avoid excessive contamination.

Nuclear radiation may seriously affect normal working of solid state devices. Components for use in systems likely to be exposed to such radiation (for example, industrial control and space vehicle systems) are often specially manufactured and treated to minimise the radiation effects.

Atomic behaviour of all materials, but particularly electrically conducting materials, depends to a large extent on material temperature. Temperatures in various geographical locations can vary from –40° C to +65° C, even at ground level. It must also be remembered that even at constant ambient temperature the internal temperature of a component is affected by operating conditions, and thus even a system unlikely to be moved geographically must contain components able to withstand a working temperature above the normal maximum of the surroundings. It has been shown that component failure rate doubles with every ten degree increase in temperature, which is a most important fact.

The importance of operating stresses set up by voltage, current, power etc. is self-evident. In general there is a maximum level of these quantities for each component, beyond which the reliability falls rapidly. Devices using static electric fields through which current should not flow (for example, capacitors, certain field-effect devices, etc.) will break down if the voltage gradient exceeds a certain value. (Note: Voltage gradient is the voltage across a component divided by the distance across which the voltage is acting.) High current levels lead to excess power dissipation, high temperatures, and subsequent breakdown as indicated earlier. An important practice used to reduce failure rate arising from operating stresses, is known as *derating*. Derating restricts the use of the component to working levels of voltage, current etc. *below* the normal working values (that is, the *rated values* given by the manufacturer). Reliability may be considerably increased by derating.

11.2. Nature of Component Faults

As stated in chapter 4, components may be divided into two types, active and passive. Active components include valves, transistors and semiconductor diodes, and similar devices; passive components include resistors, capacitors and inductors. The various types of each component were described in that chapter. In this article the nature of the faults that may occur in the various components are examined, passive components being considered first.

The fault most likely to occur in a resistor is a break in the conductive path (that is, an open circuit). The fault may develop either through continued use, continual heating and cooling of the resistor causing crystallisation and an increase in the brittleness of the material thereby making it more susceptible to mechanical shock, or through an excessive current being passed through the component. If the circuit has been working efficiently until a fault develops and a burnt resistor is apparent, it is likely that a second fault exists which caused the burn-out by allowing the flow of abnormal current. Continued use of carbon resistors may produce a form of crystallisation which changes the resistance value, rendering the fault more difficult to trace because the conductive path is not completely broken. Measurement techniques are considered in the next article. Variable resistors are liable to wear mechanically and uneven contact in various positions of the rotating arm may be produced. When in use as a volume control, for example, the uneven contact leads to *spurious signal generation* (transients) which is heard as noise superimposed on the intelligence signal. Various compounds are available commercially to improve contact and thus lengthen the working life of variable resistors. Before use, such compounds should be applied to a test material similar to that of the variable resistor or its surroundings, because certain plastics dissolve in the type of cleaning fluid used for metallic contacts. If the improvement after application of a cleaning compound is short lived the variable resistor should be changed for a new one.

A short-circuit across the resistor ends is extremely unlikely with most types of fixed resistor. If this does occur, however, the measured resistance between the ends falls to virtually zero and the component must be replaced.

Capacitors may become open-circuited, short-circuited, or 'leaky'. Leaky means that the insulation resistance of the dielectric is reduced and through currents flow. A good quality capacitor in perfect working order has a very high resistance, varying from as high as 1000 MΩ for ceramic or paper dielectrics to 0.5 MΩ for electrolytics. For an electrolytic capacitor the resistance should not be too high because this may indicate electrolyte deterioration. Electrolytic capacitors should not be stored for too long a period before use, because of the possibility of such deterioration. Methods of checking capacitors are described in the next section.

The most probable fault in inductive components such as a.f. and r.f. chokes (choking coils) and transformers is an open-circuit (as with resistors). Here again the fault may develop through long use, the conductor becoming brittle with age, or through a second fault causing excessive coil current. A second possible fault is a short-circuit from coil to frame in iron-cored coils and transformers, and this should be checked if a faulty component is suspected. This fault could result in a conductive path between, say, the primary and secondary windings of a transformer. The normal resistance between windings is very high because they are normally not connected (except in the case of the autotransformer which shares a common earth lead between primary and secondary).

A complete short-circuit between ends of a single coil is unlikely. Shorted turns resulting in reduced d.c. resistance are possible but difficult to detect, except by a slight reduction in inductive performance for the inductance is also affected. Care must be taken to distinguish between d.c. resistance of a coil (which is the opposition to direct current flow, and may vary between a few ohms for an r.f. coil and a few hundred ohms for an a.f. coil) and a.c. impedance (the opposition to alternating current). The impedance, which consists of resistance and reactance, depends on inductance and frequency, typical values being about 3000 Ω for a 10 H coil at 50 Hz and 3000 Ω for a 1 mH coil at 50 kHz.

There are a number of possible faults that may occur in valves and transistors. Considering thermionic valves first, possible faults include filament open-circuit, low emission, electrode short-circuit, filament-cathode leakage and a reduction in mechanical rigidity of the electrode structure which produces an effect called *microphony*. A break in the heater filament prevents heater current flow and thus thermionic emission. The valve is then incapable of conduction. The emissive surface of a valve cathode deteriorates with age and emission falls off. This results in reduced performance and may be suspected if static voltage tests, described in the next article, indicate a reduced valve current. An electrode short-circuit may occur if the valve has been subjected to severe mechanical shock; or, occasionally, in normal use, if the valve structure contains inferior material. The effect on the circuit depends particularly on which electrodes are involved, a short between grids of a multigrid valve, for example, having a different effect than a grid-cathode or grid-anode short. Equal voltage readings at different electrodes not normally held at the same potential may indicate internal short-circuits of this nature.

In indirectly-heated valves the heater is insulated from the cathode and extra circuit precautions to prevent interaction between heater and cathode are normally less essential than when directly heated valves are used. If, however, the insulation resistance falls off in continued use—owing to repeated heating and cooling—leakage currents may flow between cathode and heater and a mains frequency noise signal may be superimposed on the valve current. In audio amplifiers this would be noticed as a hum in the loudspeaker.

Valve microphony is vibration of the electrodes producing changes in the internal electric fields. These register as fluctuations in the valve current and in turn introduce a noise signal into the system. Whether or not the electrodes vibrate sufficiently to cause microphony is determined by the rigidity of the structure. As with the shorted electrode fault, loss of rigidity may be caused by abnormal mechanical shock or by inferior material used during manufacture. Normally, inferior material is picked out during the manufacturing process, but occasional faulty structures can escape inspection. Many valves, especially multigrid valves, produce noise set up by internal field fluctuation that is unconnected with electrode vibration. The effect is called the *shot effect* and may be apparent in the form of hissing in systems having an audio output. Certain levels of noise are tolerable, but beyond this it may be necessary to replace the valve.

In general it should be noted that, from a d.c. point of view, a valve is a one-way device with zero reverse current. This simplifies static testing, which usually gives reliable readings very little affected by meter insertion. Solid state devices do not have this advantage. This point is further considered in the next article. One exception to the general rule in valve circuits is where valve capacitance produces a charging effect and the erroneous impression may be gained that valve current is flowing when it is not. This is also considered in the next article.

Common faults in solid state devices, particularly transistors, include short-circuited electrodes, open-circuited electrodes and crystal structure damage. One or more of these faults may be caused in transistors by excessive current arising from voltage-overloading at the collector (or drain) or too high a bias at the base (or gate). In field effect transistors, in particular, damage can occur if too high a forward bias is applied to a gate-source junction that is normally reverse biased. Voltage-overloading may occur even at correct supply voltages if the transistor is subjected to transient surges. These occur when a reactive circuit is opened or closed. It is therefore not advisable to insert a transistor into a 'live' circuit because the device may then receive maximum transient voltage. Control of the circuit by conventional switching at the power supply also causes transients but their effect is reduced by the time the power reaches the transistors in other subunits. A third possible cause of fault conditions usually resulting in crystal damage, is the application of excessive heat to the transistor, and this should be borne in mind when removing transistors from assemblies of discrete components. In integrated circuits access to individual transistors is difficult and fault conditions are transmitted to the outer connections of the overall circuit. The integrated circuit is thus best treated as a subunit and fault location tests performed as described in the next article.

11.3. Fault Location

In the location of a fault within a system it is first necessary to determine which subunit is not functioning correctly. Often this may be obvious from the nature of the system and of the fault. For example, consider a domestic radio receiver giving an output with pronounced hum superimposed upon the normal signal. Here, it would be probable that the receiver mixer, i.f., detector, and a.f. stages were functioning correctly, because normal signal is present. The obvious choice for an immediate check is the power supply subunit, because it is here that pronounced hum is most likely to originate. If the power supply were found to be correct, then other subunits should be checked in the way about to be described. However, following up an immediate suspicion in this way often reduces the time spent in fault location. Even when fault conditions do not necessarily indicate the power supply to be at fault it is nevertheless advisable to check that all subunits are receiving a correct supply before proceeding with further tests.

If there is no indication which subunit is faulty a technique often used is the *half-split* method. Here the system is first divided into distinguishable subunits and a break is then made halfway along the system. Each half is then checked dynamically for signal transmission, by inserting a signal similar to the one normally present at the input of the half-system and then testing the output of the half-system. Consider the block diagram of the AM superhet receiver shown in figure 11.1. An appropriate first break is between the i.f. strip and the detector/first audio stage. The first half is tested by replacing the aerial input by a signal generator giving an amplitude modulated output at a frequency to which the mixer stage is tuned (for example, if the receiver is set to the long wave band and tuned to BBC Radio 2 the carrier frequency

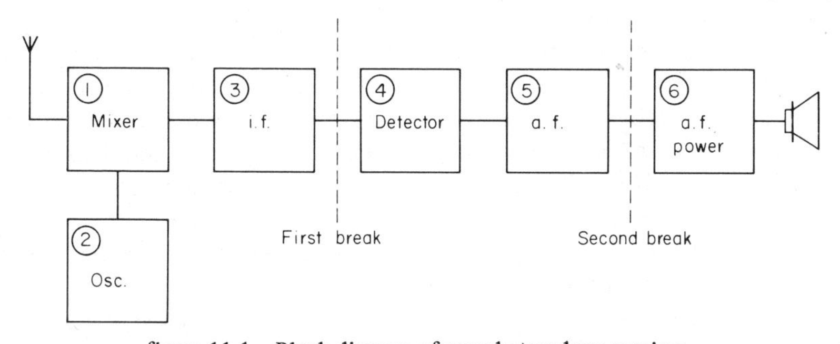

figure 11.1 Block diagram of superheterodyne receiver

should be 200 kHz). The output of the first half at the last i.f. stage before the detector should then be an amplitude modulated carrier at the frequency of the i.f. (usually given in manufacturers' data). The reason for the use of the signal generator is that a controllable amplitude signal is guaranteed to be present and can be adjusted to a suitable level. If everything else appears functional the aerial can be checked as a separate component later. (A fault occurring actually in the aerial, except for an obvious and usually visible break in the connecting wire, is unusual, because receiving aerials are not subject to high fluctuating power levels in the same way that other subunits are.)

If the first half of the system is functioning correctly the second half is now tested in a similar manner. An amplitude modulated wave at a carrier frequency equal to the i.f. is fed to the detector. If this half is functioning a tone should be heard at the loudspeaker at a frequency equal to the modulating frequency (1 or 2 kHz perhaps). The modulating frequency is set at the signal generator. If either half is not functioning correctly the half-system is split into two further halves and the process repeated. In the example given suppose the second half was faulty the half-system could be split between the detector/first audio stage and the audio output stage. Insertion of a modulated carrier into the detector should produce an audio signal at the output of the first audio stage. This may be checked using a cathode ray oscilloscope (CRO).The audio output stage may be checked by inserting a low amplitude audio frequency signal at the input and checking the loudspeaker output by ear. By progressively applying the half-split method the faulty subunit should eventually be located. This circuit may then be checked stage by stage using both static (d.c. level) tests and dynamic (signal) tests.

Active devices used in the majority of electronic circuits require a d.c. power supply and all points in the circuit are at a particular d.c. potential with respect to earth (usually taken as the lower supply line in the schematic diagram, whether or not the line is actually earthed) ranging from zero volts to the supply voltage delivered by the power supply unit. These levels of voltage for normal working are called the quiescent levels. Static testing involves measuring the voltage at the various points and comparing the readings with those given by the manufacturer or, in the event that these are not available, with levels which theory indicates would be of the correct order. If the measured levels are incorrect, logical deduction should quickly lead to the component that is causing the trouble.

An ideal voltmeter does not draw current and thus does not change existing circuit voltage levels when it is inserted to measure those levels. The electronic voltmeter (valve or transistor) is thus the best instrument for static testing. However, in practice, an electronic voltmeter is not always available and measurements are commonly made by means of a multimeter (ammeter-voltmeter-ohmmeter in one instrument) such as the 'AVO' or 'Selectest'. These instruments are of the moving-coil variety and thus require current to give an indication on the scale. The higher the instrument resistance on the volt ranges, the less current is drawn by the meter and the more reliable are the readings. Voltmeter resistance is usually expressed in 'ohms per volt', which represents the meter resistance on any one scale divided by the voltage for that scale. For example, a multimeter having a resistance given as 22 000 ohms/volt will have a total resistance on the 30 V scale of 30 x 22 kilohms, which is 660 kilohms; and this is the value of the shunt resistance that the meter introduces into the circuit when making measurements on the 30 V scale. The importance of having a high voltmeter resistance is shown in the following example. The circuit of figure 11.2 shows a potential divider circuit made up of 30 kilohms and 60 kilohms resistors in

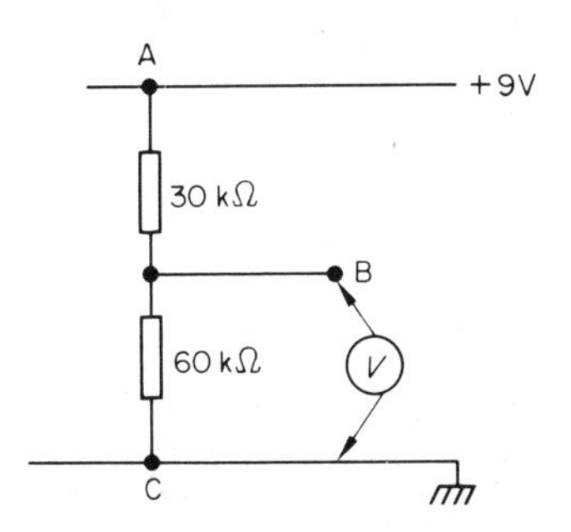

figure 11.2 Potential divider

series, the two being supplied by 9 V d.c. By calculation the p.d. between B and C should be 9(60/90) which is 6 V, if both resistors are in good order. Insertion of a 22 000 ohm/volt instrument on the 10 V range to measure this voltage puts a resistance of 220 000 ohms in parallel with the 60 k resistor. The resistance between B and C is now (220 x 60)/(220 + 60) kilohms, which is (approximately) 47 kilohms. The p.d. between B and C is now 9 x 47/(47 + 30) which is 5.5 V. Notice that even this instrument which, as was stated, is commonly used and is regarded as sufficiently accurate for most normal testing, modifies the voltage in this case by 0.5 V in 6 V which is over 8%. Consider now an inferior instrument of only 200 ohms/volt being used on the 10 V range to measure the p.d. between B and C. The meter resistance is now 2000 ohms and the resistance between B and C is 2000 ohms in parallel with 60 000 ohms, giving an equivalent resistance of (60 x 2)/(60 + 2) kilohms, which is approximately 1935 ohms. The voltage across BC is now 9 x 1935/(1935 + 30 000) volts, which is about 0.6 V. The inferior instrument has changed the voltage by 5.4 V in 6 V, which is about 90%. It is clear that the effect of the measuring instrument on the circuit during test must always be borne in mind. Another example is given later in this article.

Interpretation of the voltage measurements is aided by the application of

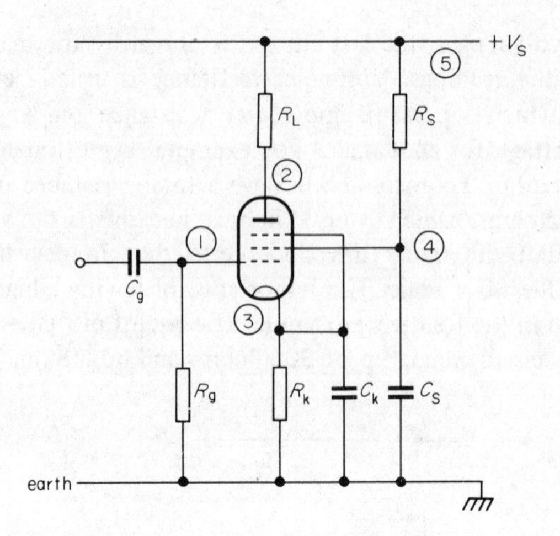

figure 11.3 Valve amplifier

normal circuit theory. For instance, consider a valve known to be part of a class-A audio amplifier using cathode bias as shown in figure 11.3. The valve shown is a pentode requiring correct screen grid voltage in addition to voltage requirements at anode, cathode and grid. Normal quiescent voltage levels at test points 1 to 5 might be as follows.

Test point	1	2	3	4	5
Voltage	0 V	240 V	5 V	200 V	300 V

(All voltages measured with respect to earth.)

If the HT line (test point 5) were at zero volts and the power supply unit (elsewhere in the system) were working correctly, a break in the line between power supply and amplifier would be indicated. In this example all test points would also register zero volts.

If the HT were present, but the anode (test point 2) were at zero volts two possibilities would exist: the anode might be short-circuited to earth, or the load resistor R_L might be open-circuit. If the first fault were present, resistor R_L would have the full HT across it, so that heavy current would flow and the resistor would be hot; while test points 3 and 4 would register abnormal voltage levels as only screen current would flow. If the second fault were present (that is, resistor R_L were open-circuit) no anode current would flow, R_L would be cold, and test points 3 and 4 would again register abnormal voltage because of the screen current acting alone.

If the cathode were at zero volts and the anode registered a much higher voltage than normal (almost at HT) the indication would be that valve emission had ceased. (Note that even when the anode voltage of a cut-off valve is at full HT, the voltmeter current flowing through R_L reduces the anode voltage meter-reading slightly.) If the cathode were found to be at zero volts, with the anode and screen at a much lower potential than normal, then the indication would be that the valve was unbiased thereby allowing a higher-than-normal anode current. The probable fault would be a short-circuit in capacitor C_K. (Notice that zero volts at the cathode does not necessarily indicate an open-circuited cathode resistor.) If R_K were open-circuit and C_K were an electrolytic capacitor, then leakage current could flow via C_K and thus through the valve. A voltage would then be developed at the cathode and this would constitute valve bias. The effective bias could in fact be higher than normal, resulting in reduced anode current and a higher anode voltage almost equal to HT. The screen voltage would also be higher than normal. Although not all possible faults in this circuit have been considered, those already examined do provide an indication of the type of reasoning that it is necessary to employ in order to deduce the location of a fault.

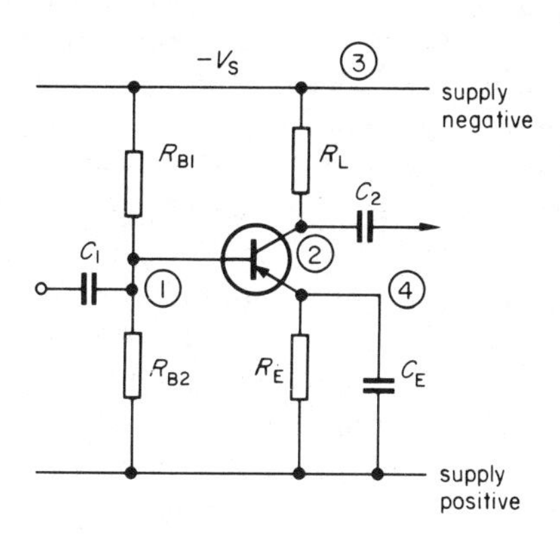

figure 11.4 Transistor amplifier

As a second example consider the temperature stabilised common-emitter amplifier of figure 11.4. Normal voltages measured between the test points indicated and the supply positive (earth) might be

Test point	1	2	3	4
Voltage	−1.6 V	−5.2 V	−10 V	−1.5 V

As with the valve circuit, if test point 3 (the supply) were at zero volts and all other test points at zero also, a break in the supply line negative would be indicated. If test point 3 were at normal voltage but the other test points were not, then the following possibilities exist.

If the collector-earth voltage (test point 2) were almost equal to the supply negative voltage and the emitter voltage were not zero, the indication would be that the collector lead inside the transistor was open. The absence of collector current might also change the p.d. across R_{B1} and the base-earth voltage (test point 1) would then differ slightly from normal. In fact, since the supply is the sum of the voltage drop against R_{B1}, the base-emitter voltage and the voltage drop across R_E (and this latter would be reduced to a very low value), then the probability would be that the voltage drop across R_{B1} would be increased, thereby reducing the base-earth voltage. In the valve circuit of figure 11.3, a break in the load resistor would cause the voltage reading at the anode (test point 2) to be zero. In the transistor circuit of figure 11.4, however, a break in the load resistor would cause the reading at the collector (test point 2) to be very high. The reason for this difference is that in the transistor circuit current can flow from earth through the moving coil meter, through the collector-base junction and thence via R_{B1} to the supply, thereby providing a voltmeter reading almost equal to the supply voltage. This fundamental difference of transistors (that is, current paths between collector and base and between emitter and base) should always be borne in mind when attempting to interpret readings.

If the base-earth voltage were zero R_{B1} might be open-circuit. The transistor would then be unbiased and the collector current would be zero. The emitter-earth voltage would be zero and there would be zero voltage drop across the load resistor. The collector-earth voltage would not be equal to the supply (as might be falsely expected in a cut off transistor) because voltmeter current would reduce the collector voltage slightly below the full supply voltage, the exact amount being dependent upon voltmeter resistance.

If the emitter-earth voltage were zero but the base-earth voltage were not, the possibility of an open-circuited emitter would exist; in which eventuality, emitter-base current would cease, thereby reducing the voltage drop across R_{B1} (as R_{B1} normally carries the base current as well as the bleed current of R_{B1} and R_{B2}). Thus the base-earth voltage, which is the difference between the supply voltage and the p.d. across R_{B1}, would rise to a value somewhat above that of normal forward bias. A second possibility with zero emitter-earth voltage would be that a short circuit might develop in capacitor C_E, thus effectively removing R_E from the circuit and replacing it by zero resistance. Transistor bias (which is the difference between base-earth and emitter-earth voltages) would increase considerably and large base and collector currents would flow. The base-earth voltage would thereby be reduced as a result of the increased voltage drop across R_{B1} and the transistor might even go into saturation (indicated by very low collector-earth voltage and a large voltage drop across transistor load R_L), which may damage the transistor if the current is large enough. Notice that an open-circuited emitter resistor R_E would not necessarily result in zero emitter-earth voltage, because the existence of a current path between emitter and base could give a voltage reading (on a moving coil meter) taken between emitter and earth almost equal to that between base and earth. With R_E open circuit the voltage readings would in fact be similar to those with an open-circuited emitter: no base current, an increased base-earth voltage, and a collector-earth voltage almost equal to the supply. However the emitter-earth voltage would not be zero.

Another possible fault in this circuit could be an open-circuited base. This would effectively cut off the transistor (except perhaps for the collector-emitter leakage current), and would give a high collector-earth p.d. together with an increased p.d. between base and earth arising from the absence of base current.

Not all faults have been considered but, as with the valve circuit, sufficient indication has been given of the logical reasoning that should be employed in fault finding.

11.4. Component Testing

Once a particular component is suspected of being faulty, individual tests must then be performed. Ideally, the component should be removed from the circuit completely; but, if this is inconvenient (at least until it proves necessary for removal), one or more leads should be disconnected and care taken to avoid current paths in neighbouring components when testing. Such paths may give false readings and lead to incorrect conclusions. Methods for testing components are described below.

Resistors are normally checked with an ohmmeter (in all probability on one of the resistance ranges of a multimeter). Such an instrument carries its own power supply and the circuit under test must be disconnected from the subunit power supply if the resistor is only partially removed from the circuit (that is one end disconnected). Zero resistance on an ohmmeter is normally full scale deflection of the pointer and care must be taken not to confuse this reading with 'infinite ohms'. With the meter leads connected together, the ohmmeter is first zeroed, using the electrical control provided. This removes lead resistance from the reading and adjusts for any deterioration of battery voltage. If the instrument is a multirange meter, take care to adjust the zero control appropriate to the chosen range.

In selecting the appropriate resistance range the approximate value of the resistance to be measured must be borne in mind. A 5-MΩ resistor, for example, will indicate infinite resistance on a 1-MΩ scale; and an open circuit may then be suspected, though in fact it does not exist. Similarly, a 100-ohm resistor may read zero on a 1-MΩ scale; a more appropriate scale would be, say, 1000 ohms, or 10 000 ohms (depending on what is available). When checking high resistance it is advisable not to touch the lead connections because the human body resistance, which may be as low as 50 000 ohms, may adversely affect the reading.

Capacitors may also be checked for component resistance by use of an ohmmeter. On connection the meter initially reads low; then, if the capacitor is functional, the pointer moves to the high resistance end of the scale as the component charges. The reading given when the pointer stops moving is the insulation resistance, which is normally high if the capacitor is in good condition. Low resistance indicates a short-circuit or a leaky capacitor; very high resistance indicated immediately (that is, without charging) may indicate an open-circuit except for very low-value capacitors in which the charging time is too short to cause detectable pointer movement as described. Capacitance itself may be measured on a capacitor bridge; the instructions for use of these instruments depend upon the type and are usually given with the instrument.

Inductors may be checked with an ohmmeter in the manner described for resistors, bearing in mind that inductor d.c. resistance is usually low. Inductance itself may be measured either by using a reliable a.c. supply to determine the inductive reactance or by a direct reading from a bridge instrument constructed for the purpose.

Valves may be tested with an instrument built for the purpose, taking care to follow closely the manufacturer's instructions. Quick check instruments are available which indicate the valve performance in one of the three categories 'good', 'bad' or 'doubtful'. For a more detailed analysis of performance a more sophisticated instrument is required.

Transistors may also be tested using one of the many checking instruments available and the same comments apply here as to valve testers. However, an indication of whether or not a transistor is functional may also be obtained by use of an ohmmeter on a high resistance range (a high range is chosen to avoid applying excessive forward bias to the device). The forward and reverse resistance of the transistor collector-base junction and emitter-base junction are measured using the meter leads connected first one way, then reversed. The reverse resistance is normally about ten thousand times the forward resistance. Actual values depend, of course, on the transistor, but forward resistance is usually a few hundred ohms and reverse resistance a few hundred thousand ohms for germanium devices and perhaps one or two million ohms for silicon devices. Solid state diodes may also be checked in this manner. If a transistor or diode of the same type is available and is known to be in good working order, it is useful to check similarly taken readings for each device. If the readings are of similar magnitude the component may then be assumed to be functional. An appropriate test instrument must, of course, be used if more detailed analysis is required.

Appendix 1
Resistor colour code and schematic diagram code BS 1852

The resistance of most fixed resistors (other than wirewound types) is usually indicated by the use of a system of coloured rings grouped together at one end of the component. There are usually either three or four of these rings and they are read as follows: rings one and two, counting from the end at which the rings are grouped, give the first two figures of the resistance value; ring three gives the number of noughts following these two figures; and ring four (if present) gives the tolerance (that is, a measure of the range within which the actual resistance is permitted to lie). The colour code (table A1.1) is as follows:

Colour	Black	Brown	Red	Orange	Yellow	Green
Figure (or number of noughts, for third ring)	0	1	2	3	4	5

Colour	Blue	Violet	Grey	White	Silver	Gold
Figure (or number of noughts, for third ring)	6	7	8	9	–	–

table A1.1

If a fourth ring is present it will be silver to indicate 10% tolerance or gold to indicate 5% tolerance. If a fourth ring is not present the tolerance may be assumed to be 20%.

Thus, a resistor coloured red-red-red-silver has a nominal resistance 2200 (that is, 2–2–two noughts) ohms and its actual resistance lies within ±10% of

2200, that is between (2200 – 220) ohms and (2200 + 220) ohms. Similarly, a resistor coloured yellow-violet-brown is 470 ohms, 20% tolerance. A third ring coloured black signifies no nought so that a brown-black-black set of rings would indicate 10 ohms, that is first two figures 1 and 0, no noughts to follow.

BS 1852 Resistance Code

Throughout the book schematic circuit diagrams have indicated resistance values using standard prefixes. The new British Standard 1852 Code is being adopted by some manufacturers. This code tells more about the resistor but uses fewer characters. Some examples are

0.56 Ω is written R56
1.0 Ω is written 1R0
5.6 Ω is written 5R6
68 Ω is written 68R
2.2 kΩ is written 2K2
10 MΩ is written 10M

An additional letter may then be used to give the tolerance.

F = ± 1%
G = ± 2%
J = ± 5%
K = ± 10%
M = ± 20%

So that

4R7K means 4.7 Ω ± 10%
68KK means 68 kΩ ± 10%
4M7M means 4.7 MΩ ± 20%
6K8J means 6.8 kΩ ± 5%

and so on.

Appendix 2
Matching

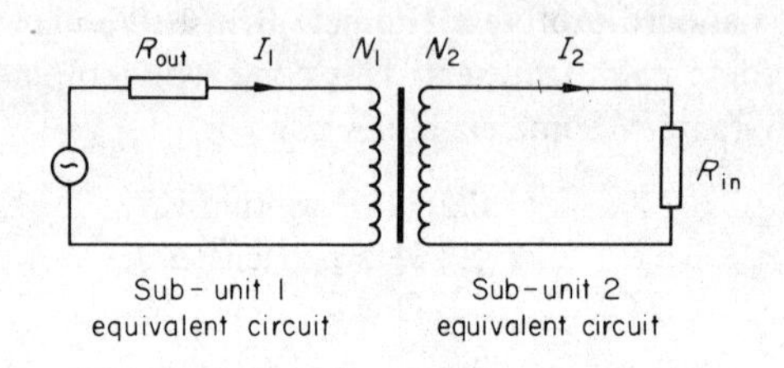

figure A2.1 **Power transfer**

Maximum power is transferred from one subunit to the next (see figure A2.1) when the *output resistance* of one is equal to the *input resistance* of the next. When this condition is not met, a transformer is often used between subunits because of its ability to 'reflect' impedance. In the circuit shown which indicates the equivalent output circuit of one subunit coupled to the equivalent input circuit of the next via a transformer, the power in the input resistance R_{in} is given by $I_2^2 R_{in}$ and the power in the output resistance R_{out} is given by $I_1^2 R_{out}$. Maximum power is transferred when

$$I_2^2 R_{in} = I_1^2 R_{out}$$

or when

$$(R_{in}/R_{out}) = (I_1/I_2)^2.$$

But $(I_1/I_2) = (N_2/N_1)$, where N_2/N_1 is the ratio of secondary turns to primary turns on the transformer (see chapter 4). Thus

$$(R_{in}/R_{out}) = (N_2/N_1)^2;$$

so if the ratio N_2/N_1 is *chosen* to equal $\sqrt{(R_{in}/R_{out})}$, this equation is satisfied and maximum power is transferred. The transformer is then said to 'reflect' an apparent resistance equal to the output resistance from subunit 2 to subunit 1.

Appendix 3
Valve and transistor parameters

A parameter is a characteristic or property of a valve or transistor which identifies it and distinguishes it from another similar valve or transistor. In choosing a particular transistor or valve to do a specific job, a comparison of parameters is the means by which the most suitable component can be selected. The important parameters of a valve or transistor are the same as those for any amplifier, namely, input and output impedances, gain and feedback ratios. The amplifier characteristics will, of course, be determined by the parameters of the active devices it contains.

Parameters can be expressed in the units of impedance or resistance, in the units of admittance or conductance, or they may have no units and be just plain numerical ratios. The same kind of unit (that is, ohms or siemens) may be used for all four parameters, or the parameters may have mixed units (for example, ohms, siemens and straight ratio). To clarify this, consider expressing the four parameters in ohms; the input and output impedances are already in ohms; the gain and feedback would have to be expressed in the form of a voltage change divided by a current change, because the unit of voltage divided by current is the ohm. Expressing a gain in this form will not necessarily suit the device. A triode valve, for example, has a current change produced by a voltage change. The unit of current divided by voltage is the reciprocal of the ohm, the siemens. Without going further into detail, it is sufficient to say that appropriate parameters are chosen for a device, depending upon how the device works. Thus there are different parameters used for valves than for transistors and again different parameters for the bipolar device than for the unipolar device.

Parameters expressed in ohms are called z parameters at high frequencies and r parameters at low frequencies. Parameters expressed in siemens are called y parameters at high frequencies and g parameters at low frequencies. The parameters used for each device will now be given.

Of the four parameters normally of interest, only two are used for valves, one of these being expressed in two ways. The two not applicable (at least at low frequencies) are input impedance, which is normally assumed to be very high (because no current flows), and the feedback ratio, which (again at low frequencies) is assumed to be zero. The forward transfer ratio, which is a measure of the available gain, is expressed either in siemens or as a ratio.

The *mutual conductance,* g_m, is the ratio of a small change in anode current to the small change in grid voltage which causes it. Thus, for a valve in which a *change* in grid voltage of 1 V causes a *change* in anode current of 5 mA, the mutual conductance is 5 mA/V or 5 millisiemens. The mutual conductance is actually the *slope* of the I_A/V_G curve (the mutual characteristic) shown in figure 4.17b. (See also figure 6.3a.)

The *amplification factor,* μ, is the ratio of the change in anode voltage to the change in grid voltage for the same change in anode current. For example, if a 1 V grid change causes a 5 mA anode current change and a 20 V anode voltage change also causes a 5 mA anode current change, the amplification factor is 20. It has no units because it is one voltage change divided by another. Both μ and g_m will affect the overall amplifier gain when the valve is put to use.

The third parameter is the anode resistance, r_a, which is the output resistance of the valve. It is defined as the small change in anode voltage divided by the resultant change in anode current when the grid voltage is constant. For example, if a 20 V anode voltage change causes a 5 mA anode current change when the grid voltage is constant, the anode resistance is 20/5 kilohms. The anode resistance is, in fact, the *reciprocal* of the slope of the I_A/V_A curves shown in figure 4.17a for a triode, and 4.25 for a pentode. The slope of the pentode curves is very small, thus the output resistance, r_a is very high. See figure 6.3b.

The three valve parameters are related to one another, as the following expression indicates.

$$\frac{(\text{small change in } V_A)}{(\text{small change in } I_A)} \times \frac{(\text{small change in } I_A)}{(\text{small change in } V_G)} = \frac{\text{small change in } V_A}{\text{small change in } V_G}$$

which may also be written

$$r_a \times g_m = \mu$$

Thus, if any two parameters are known, the third may be found.

The valve parameters r_a, g_m and μ, are measures of output resistance, mutual (transfer) conductance and amplification, only when the input is the control grid and the output is the anode (that is, for a common-cathode amplifier). Modifications of these parameters are used for the other modes, common-grid and common-anode. Special parameters for these modes are not usually defined for the valve, though their equivalents are specified for the transistor.

The most commonly used parameters for bipolar transistors are the hybrid, or 'h', parameters. The 'h' parameters are input resistance, h_i; output conductance, h_o; current gain, h_f; and feedback ratio, h_r. The input resistance is not extremely high, because for a bipolar transistor current flows at the input. The output resistance, if required, is the reciprocal of h_o; but the output conductance is usually given because it is a more suitable parameter for this particular device. As before, the parameters are defined using small changes so that

$$h_i = \frac{\text{small change in input voltage}}{\text{small change in input current}}$$

$$h_o = \frac{\text{small change in output current}}{\text{small change in output voltage}}$$

$$h_f = \frac{\text{small change in output current}}{\text{small change in input current}}$$

$$h_r = \frac{\text{small change in feedback voltage}}{\text{small change in output voltage}}$$

The parameters h_f and h_i are measured with the output terminals short-circuited (that is, with no load); h_o and h_r are measured with output terminals open-circuited (that is, with zero resistance load). An examination of the definitions shows that h_i is measured in ohms, h_o in siemens, and that h_f and h_r are pure numbers. Because of the mixed units, these parameters are called *hybrid parameters,* the 'h' standing for hybrid. As with valve parameters, the 'h' parameters of a transistor can be obtained from the characteristics. For example, if a graph of collector current against base current is plotted (see figure 6.3c) the slope would represent h_f for the mode where output current is collector current and input current is base current (that is, the common emitter mode).

For the transistor (unlike the valve), a set of 'h' parameters is often given for each mode; for common-emitter, common-base and common-collector. The basic definitions remain the same, but the output and input voltages are those at the output and input electrodes for the particular mode. To distinguish between modes a second suffix is used in which 'e' represents common emitter, 'b' common base, and 'c' common collector.

Thus, h_{fe} which is the ratio between a small change in output current to a small change in input current for the common emitter mode means, in fact, the ratio between a small change in collector current (the output current in this case) and a small change in base current (the input current in this case). Similarly, h_{fb} is the ratio between small changes of collector current (output) and emitter current (input) for the *common base* amplifier. Other symbols used for h_{fe} and h_{fb} are β and α respectively, but these symbols are falling out of use at the present time. The use of capital letters in the suffix, for example, h_{FE} h_{FB}, etc. removes the words 'small change' from the definition and the parameters apply to unchanging quantities (that is, d.c.). Thus, where h_{fe} means a small change in collector current divided by small change in base current, h_{FE} means the d.c. value of collector current divided by the d.c. value of base current. The parameters h_{fe}, h_{fb} etc. are close in value to h_{FE}, h_{FB} etc. if h_{fe}, h_{fb} are measured at small signal levels. Because transistor parameters are rather more complex than valve parameters, they are summarised and compared in table 3A.1.

The most convenient parameters for field-effect transistors are the '*g*' parameters, all of which are measured in the units of conductance (siemens). The four *g* parameters are very similar to the *h* parameters for bipolar transistors. The *g* parameters are

$$\text{input conductance } g_i = \frac{\text{small change in input current}}{\text{small change in input voltage}}$$

$$\text{output conductance } g_o = \frac{\text{small change in output current}}{\text{small change in output voltage}}$$

$$\text{transfer conductance } g_f = \frac{\text{small change in output current}}{\text{small change in input voltage}}$$

$$\text{feedback conductance } g_r = \frac{\text{small change in feedback current}}{\text{small change in output voltage}}$$

Notice that all the parameters are measured in terms of current/voltage ratios, and the unit will therefore be the siemens as stated. As with *h* parameters, a second suffix is used to show the mode; *s* for common-source; *g* for common-gate; and *d* for common-drain. Thus g_{fs}, for example represents a small change in drain current divided by a small change in gate voltage (for the common-source connection, the drain current is the output current and gate voltage is the input voltage). The parameter g_{fs} is the equivalent parameter for the FET to g_m, the mutual conductance for a valve. As with the valve, it is obtainable from the characteristics, being the slope of the I_D/V_G curve (see figure 6.3d). Another similarity with the valve is that the input conductance is taken as zero (that is, an *extremely high input resistance*) and the feedback ratio g_r is ignored at low frequencies. The remaining parameters for the common source connection g_{fs} and g_{os} are similar to g_m and $1/r_a$ for the valve. As with the bipolar transistor, capital letters in the suffix denote the d.c. parameter, g_{FS}, for example, being the ratio of I_D to V_G (both d.c. values). See table 3A.1 for an overall summary of parameters.

	Bipolar transistors				Unipolar transistors (FET)				Valves	
	Hybrid (h) parameters	Common base	Common emitter	Common collector	Conductance (g) parameters	Common gate	Common source	Common drain		Common cathode, grid, or anode
Input parameter	small change in input voltage / small change in input current (input impedance)	h_{ib}	h_{ie}	h_{ic}	small change in input current / small change in input voltage (input conductance)	g_{ig}	g_{is}	g_{id}		
Output parameter	small change in output current / small change in output voltage (output conductance)	h_{ob}	h_{oe}	h_{oc}	small change in output current / small change in output voltage (output conductance)	g_{og}	g_{os}	g_{od}	small change in output voltage / small change in output current (output resistance)	r_a
Forward transfer (gain) parameter	small change in output current / small change in input current (current gain)	h_{fb}	h_{fe}	h_{fc}	small change in output current / small change in input voltage (transfer or mutual conductance)	g_{fg}	g_{fs}	g_{fd}	small change in output current / small change in input voltage (mutual conductance) and also	g_m
Reverse transfer (internal feedback) parameter	small change in feedback voltage / small change in input voltage	h_{rb}	h_{re}	h_{rc}	small change in feedback current / small change in output voltage	g_{rg}	g_{rs}	g_{rd}	small change in output voltage / small change in input voltage (amplification factor) N.B. μ is also a forward transfer parameter as is g_m.	μ

Note: Capital letter suffix means d.c. quantities.

table 3A.1: Active device parameters

Appendix 4
Multiple and submultiple units

As indicated in the text, units of resistance, capacitance and inductance are ohms, farads and henrys respectively. These are abbreviated Ω, F and H. The single unit is not always a convenient size and accordingly multiple units or sub-multiple units are used. The abbreviations and symbols in use are

pico	p	10^{-12}
nano	n	10^{-9}
micro	μ	10^{-6}
milli	m	10^{-3}
centi	c	10^{-2}
kilo	k	10^{3}
mega	M	10^{6}
giga	G	10^{9}

Thus: 1 gigahertz (1 GHz) means 1×10^9 or 1000 million hertz
1 kilohm (1 kΩ) means 1000 ohms
1 microampere (1 μA) means 1×10^{-6} amperes; that is, one millionth of one ampere
1 nanofarad (1 nF) means 1×10^{-9} or 1/1 000 000 000 of one farad

and so on.

Self-test questions

1. The main disadvantage of d.c. signal transmission is

A. Only low power signals may be transmitted
B. D.C. power supplies are required
C. Electromagnetic propagation cannot be used
D. The extent of the transmitted intelligence is limited

2. The synchronising signal in a television transmission is used to

A. Separate sound from vision signals
B. Ensure line and frame timebases run at the same frequency
C. Ensure camera and receiver timebases run in synchronism
D. Ensure the same brightness at the camera and at the receiver

3. A superheterodyne receiver system

A. Has poor selectivity
B. Uses a frequency changer
C. Has poor sensitivity
D. Uses wide band amplifiers

4. The purpose of the screen grid of a tetrode valve is to

A. Reduce secondary emission
B. Reduce the effect of the anode-grid capacitance
C. Provide a second signal input electrode
D. Increase the maximum possible value of anode current

5. For correct working of an NPN bipolar transistor the electrodes named should be at the following polarities with respect to the emitter

A. Collector positive, base negative
B. Collector negative, base positive
C. Collector positive, base positive
D. Collector negative, base negative

6. The nature of the impedance of a series tuned circuit below and above resonance respectively is

A. Inductive, capacitive
B. Capacitive, inductive
C. Capacitive, resistive
D. Resistive, inductive

7. The nature of the impedance of a parallel tuned circuit below and above resonance respectively is

A. Inductive, capacitive
B. Inductive, resistive
C. Capacitive, inductive
D. Resistive, capacitive

8. If the gate-source voltage of a junction gate N-channel field effect transistor is positive, that is gate is positive with respect to source

A. The drain current increases
B. The drain current is reduced
C. The drain current ceases
D. The drain current remains at the same value as when the gate-source voltage is zero

9. , For normal operation of a pentode valve the polarity of the voltages at the electrodes named (with respect to the cathode) is

A. Anode positive, screen grid positive, control grid negative
B. Anode positive, screen grid zero, control grid negative
C. Anode negative, screen grid positive, control grid zero
D. Anode positive, screen and control grid zero

10. A thyristor may be switched off

A. By reducing the gate voltage
B. By reducing the anode voltage
C. By increasing the gate voltage
D. By increasing the anode voltage

11. If one diode of a bridge rectifier failed

A. The output would fall to zero
B. The output would be reduced full wave rectified d.c.

C. The output would be half wave rectified d.c.
D. The output would be unrectified a.c.

12. A Zener diode

A. Is used in the forward biased mode
B. Employs avalanche breakdown when reverse biased
C. May be used as a current stabiliser
D. May be used as a voltage stabiliser

13. The purpose of a bleeder resistor at the output of a power supply is to

A. Stabilise the output voltage
B. Prevent the output voltage rising to too high a value on no load
C. Reduce output ripple
D. Rectify the a.c. input

14. The voltage regulator in a power supply (a.c. to d.c.) is included

A. To remove ripple
B. To rectify the a.c.
C. To stabilise the output voltage
D. To stabilise the input voltage

15. The dynamic mutual characteristic of a valve is a graph of
A. Anode current/grid voltage, anode voltage constant
B. Anode current/grid voltage, anode voltage varying
C. Anode current/anode voltage, grid voltage constant
D. Anode current/anode voltage, grid voltage varying

16. A valve is biased class-A when the bias voltage is

A. Zero
B. At cut off
C. Beyond cut off
D. Between zero and cut off and in the linear part of the I_A/V_A curve

17. A common-emitter amplifier has the input and output signals at the following electrodes

A. Input-base, output-emitter
B. Input-base, output-collector
C. Input-emitter, output-collector
D. Input-emitter, output-base

18. A complementary symmetry circuit uses

A. Two PNP transistors
B. Two NPN transistors
C. Two triode valves
D. One PNP transistor, one NPN transistor

19. A phase splitter circuit is used

A. To provide feedback in an oscillator
B. To provide antiphase signals from a single input
C. To improve the gain-frequency response of an amplifier
D. As a detector for FM signals

20. An emitter follower has

A. Low input impedance, low output impedance
B. Low input impedance, high output impedance
C. High input impedance, low output impedance
D. High input impedance, high output impedance

21. An amplifier delivers 100 mV output when the input is 10 mV. If 1 mV is fed back so as to oppose part of the input the new overall gain of the amplifier is approximately

A. 10
B. 11
C. 9
D. 100

22. If the cathode decoupling capacitor in a cathode bias circuit were to develop an open-circuit

A. The valve bias would rise
B. The valve bias would fall
C. The amplifier gain would rise
D. The amplifier gain would fall

23. The static mutual characteristic of a triode valve is a graph of

A. Anode voltage/anode current, grid voltage constant
B. Anode current/grid voltage, anode voltage constant
C. Anode voltage/grid voltage, anode current constant
D. Anode voltage/cathode voltage, grid voltage constant

24. The fall off in gain of a transformer-coupled voltage amplifier at low frequencies is due to

A. The high reactance of the load
B. The low reactance of the load
C. The high reactance of the output shunt capacitance
D. The low reactance of the output shunt capacitance

25. The fall-off in gain of a resistance-capacitance coupled amplfier at low frequencies is due to

A. The low reactance of the coupling capacitor
B. The high reactance of the coupling capacitor
C. The low reactance of the output shunt capacitance
D. The high reactance of the output shunt capacitance

26. The fall off in gain of a resistance coupled amplifier when the signal frequency is high is due to

A. The high reactance of the coupling capacitor
B. The low reactance of the coupling capacitor
C. The high reactance of the shunt capacitance across the output
D. The low reactance of the shunt capacitance across the output

27. If the load of a resistance loaded amplifier is increased in value, the HT remaining the same

A. The load line slope is increased
B. The load line slope is reduced
C. The load line slope remains the same, the load line moving up the I_A/V_A axes
D. The load line slope remains the same, the load line moving down the I_A/V_A axes towards zero

28. A tuned-anode tuned-grid oscillator

A. Has two mutually coupled tuned circuits
B. Has two tuned circuits at the same resonant frequency
C. Uses valve capacitance to provide feedback
D. Has no tuned circuits

29. In the normal operation of an astable multivibrator

A. There is one stable state

B. There are two stable states
C. There are no stable states
D. The output is sinusoidal

30. In a Colpitts oscillator

A. One side of the tuned circuit contains a tapped coil
B. One side of the tuned circuit contains two capacitors series connected
C. Feedback is effected via a transformer
D. There is no tuned circuit

31. In a phase shift RC oscillator

A. The tuned circuit is transformer coupled
B. Phaseshift is effected by an RC network
C. Feedback is from cathode to grid (or emitter to base)
D. The frequency is controlled by varying a coil

32. In a Hartley oscillator

A. One side of the tuned circuit contains a tapped coil
B. One side of the tuned circuit contains two capacitors series connected
C. Feedback is effected by a transformer
D. There is no tuned circuit

33. An anode-bend detector for AM signals

A. Reduces selectivity
B. Operates at zero grid volts
C. Operates at cut off
D. Uses automatic bias

34. A diode detector for AM signals gives

A. Poor selectivity, good sensitivity
B. Good selectivity, poor sensitivity
C. Good selectivity, good sensitivity
D. Poor selectivity, poor sensitivity

35. An AND gate gives an output 1 if

A. At least one input is 1
B. At least one input is 0

C. All inputs are 0
D. All inputs are 1

36. With all inputs at 1

A. An AND gate gives an output 0
B. An inclusive OR gate gives an output 0
C. A NAND gate gives an output 1
D. A NOR gate gives an ouptut 0

37. Two NOR gates connected in cascade give an equivalent function to one

A. NOR gate
B. AND gate
C. OR gate
D. NAND gate

38. An OR gate gives an output of 0 if

A. At least one input is 1
B. At least one input is 0
C. All inputs are 1
D. All inputs are 0

39. A two input logic gate produces an output 0 when both inputs are 1. The gate could be

A. An OR gate
B. An AND gate
C. An inverter
D. An exclusive OR gate

40. In a 3-element Yagi aerial with one director and one reflector

A. All elements are the same length
B. The director is longer than the reflector
C. The director is shorter than the reflector
D. The director and reflector are of equal length and shorter than the dipole element

41. A piece of transmission line less than one quarter wave length long is open-circuited at both ends. It behaves as

A. A capacitive reactance
B. An inductive reactance
C. A low impedance acceptor circuit
D. A high impedance rejector circuit

42. Two NAND gates connected in cascade give an equivalent function to one

A. NOR gate
B. AND gate
C. OR gate
D. NAND gate

43. A ripple-through counter

A. Has negligible propagation delay
B. Is a synchronous counter
C. Is an asynchronous counter
D. Is not likely to cause sampling error

44. In figure Q1, if direct voltage readings between base and ground and between emitter and ground were equal, the fault could be

A. R1 open-circuit
B. C3 short-circuit
C. Transistor internal base-emitter short
D. R3 open-circuit

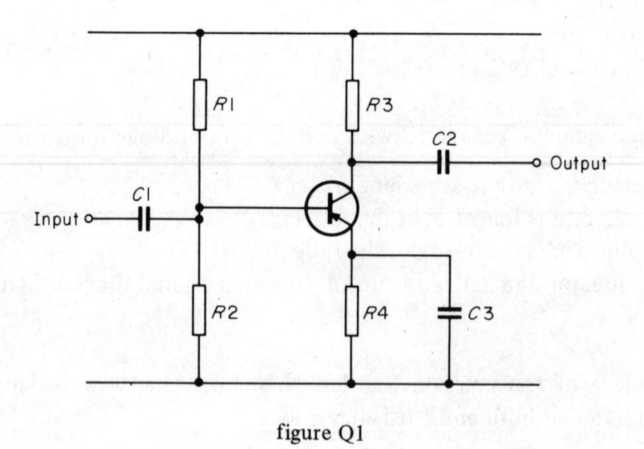

figure Q1

45. In figure Q1, if R3 went open-circuit

A. The amplifier gain would fall to zero
B. The transistor would overheat
C. The output would be distorted
D. The transistor bias would increase

46. In figure Q1 if capacitor C3 went open-circuit

A. The gain of the amplifier would be reduced
B. The transistor would cut off
C. The transistor would burn out
D. The output would fall to zero

47. In the circuit of figure Q2, if C2 went open-circuit

A. The amplifier gain would reduce
B. The output would be zero
C. The valve bias would change
D. The valve would cut off

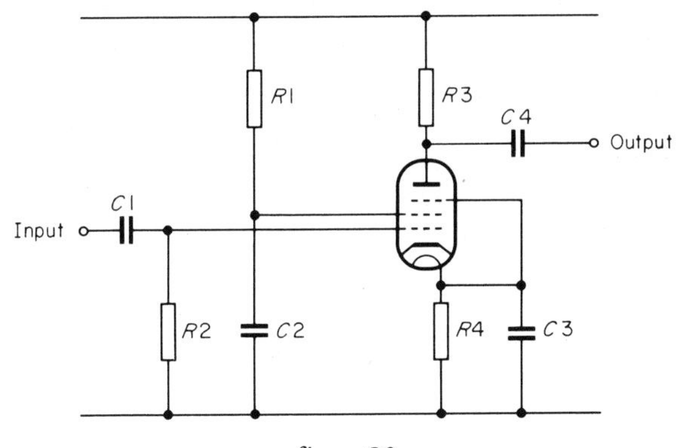

figure Q2

48. If the circuit of figure Q2 were the second stage of a two-stage amplifier and capacitor C1 went short-circuit

A. The current in R2 would fall
B. The valve would be damaged
C. The valve bias would remain unchanged
D. The output would be grossly distorted

49. In the circuit of figure Q3, if capacitor C4 went short-circuit

A. AGC would cease
B. Valve gain would fall off
C. The output would fall to zero
D. Valve bias would change

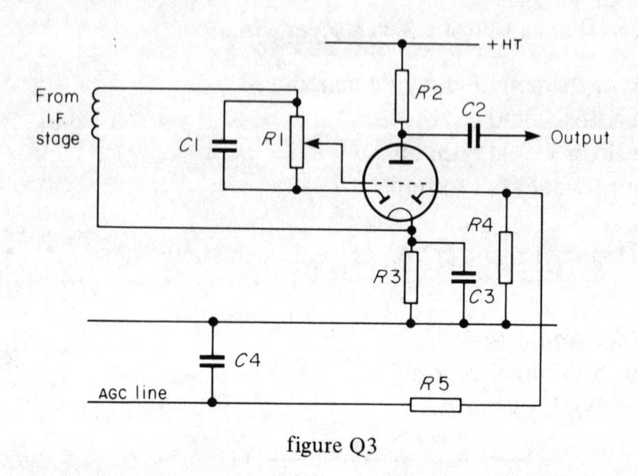

figure Q3

50. In the circuit of figure Q3 if capacitor C1 went open-circuit

A. The a.f. output would be reduced
B. The output r.f. content would increase
C. The volume control would fail
D. The AGC line would fail

Answers

1. C	11. C	21. C	31. B	41. A
2. C	12. D	22. D	32. A	42. B
3. B	13. B	23. B	33. C	43. C
4. B	14. C	24. B	34. D	44. C
5. C	15. B	25. B	35. D	45. A
6. B	16. D	26. D	36. D	46. A
7. A	17. B	27. B	37. C	47. A
8. A	18. D	28. C	38. D	48. B
9. A	19. B	29. C	39. D	49. A
10. B	20. C	30. B	40. C	50. B

Index